# THE
# NATURAL HISTORY
# OF MAN

THE

# NATURAL HISTORY

OF

# MAN;

COMPRISING

INQUIRIES INTO THE MODIFYING INFLUENCE OF
PHYSICAL AND MORAL AGENCIES
ON THE DIFFERENT TRIBES OF THE HUMAN FAMILY.

BY
**JAMES COWLES PRICHARD,** M.D. F.R.S. M.R.I.A.

*Fourth Edition, Edited and Enlarged by*
**EDWIN NORRIS,**

IN TWO VOLUMES

Vol. I.

GYAN

*Published by*

GYAN

Gyan Publishing House
5, Ansari Road
Daryaganj, New Delhi-110002
Phone: 011-47034999, 9811692060
E-mail: books@gyanbooks.com

*Distribution Network*
gyanbooks.com
India, USA, Canada, UK, Australia, France



ISBN : 978-81-212-6664-2 (Set)
ISBN : 978-81-212-6663-5 (HB)
First Published, 1855

2nd Impression 2022

*Printed at:* Gyan Press, Delhi.



**THE NATURAL HISTORY OF MAN (VOL. I)**
Author: JAMES COWLES PRICHARD

# THE NATURAL HISTORY OF MAN;

COMPRISING

INQUIRIES INTO THE MODIFYING INFLUENCE OF PHYSICAL AND MORAL AGENCIES ON THE DIFFERENT TRIBES OF THE HUMAN FAMILY.

BY

JAMES COWLES PRICHARD, M.D. F.R.S. M.R.I.A.

PRESIDENT OF THE ETHNOLOGICAL SOCIETY,
CORRESPONDING MEMBER OF THE NATIONAL INSTITUTE,
AND OF THE ROYAL ACADEMY OF MEDICINE, AND OF THE STATISTICAL SOCIETY, OF FRANCE,
MEMBER OF THE AMERICAN PHILOSOPHICAL SOCIETY; ETC.

Fourth Edition, Edited and Enlarged by

EDWIN NORRIS,

OF THE ROYAL BRITAINSOCIETY OF GREAT BRITAIN AND IRELAND.

ILLUSTRATED WITH SIXTY TWO COLOURED PLATES ENGRAVED ON STEEL, AND ONE HUNDRED ENGRAVINGS ON WOOD.

---

IN TWO VOLUMES.

VOL. I

1855.

# CONTENTS.



## BOOK I.

### GENERAL CONSIDERATIONS.

#### CHAPTER I.



#### CHAPTER II.



#### CHAPTER III.



#### CHAPTER IV.



#### CHAPTER V.



#### CHAPTER VI.

# BOOK II.

## DIVISION INTO GROUPS.

CHAPTER X.



CHAPTER X.*



CHAPTER XI.



CHAPTER XII.



CHAPTER XIII.



CHAPTER XIV.



CHAPTER XV.



CHAPTER XVI.

# INDEX TO THE ENGRAVINGS ON WOOD.

# EDITOR'S PREFACE.

The present edition of the "Natural History of Man" contains the whole of the work published by Dr. Prichard in the year 1848, together with much additional information obtained subsequently. The ruder divisions of the Human Race have, within these few years, received so much attention,—the courage of scientific travellers, the zeal of missionaries, the enterprise of traders, and the intelligent labours of government functionaries from Europe and America, have so largely contributed to our knowledge of their habits, creeds, and physiology,—that, in order to keep the book within a convenient size, it has been found necessary to take but sparingly from the sources thus placed at our disposal. In the selection made, the Editor has been guided chiefly by a wish to bring forward facts more immediately bearing upon the views which the Author himself took of his work, and in subservience to this, to record notices of any portions of the human family but little known at the time when he wrote, selecting preferably from publications not likely to be extensively diffused,—such as tracts printed at remote colonies, reports of missionaries, or bulky official papers.

That part of Dr. Prichard's work which, in the last

edition, was published in the form of an Appendix to preceding editions, is here incorporated with the rest. In most instances the additional matter supplied by him is confirmatory of the statements in former editions, and it is inserted in the place to which it belongs, just as the Author would have placed it himself, had the whole been recast; but in the few instances where, in consequence of new information obtained by the Author, the Appendix was really the substitution of a new chapter, including everything contained in the older one, this last has been omitted. The correction of obvious oversights will not be considered a departure from the original; and the Editor has not scrupled to make the spelling of proper names uniform, where the Author may have inadvertently written the same name in various ways. With these exceptions the text is rigidly adhered to: the Editor has done no more for the author than he believes he would have done for himself,—hardly so much.

All additional matter contributed by the Editor is inserted in the text, so that it may be read in connection with the subject to which it may refer; such additions are distinguished by being in brackets. Some foot-notes occasionally made upon the Author's statements are pointed out as additional.

The series of Ethnographical Maps prepared under the superintendence of Dr. Prichard to illustrate his History, will be found better adapted for reference by having the leading geographical names written in, the names of newly-discovered tribes carefully inserted, and some of those which were incorrectly located now put in their true places.

The Editor concludes by gratefully acknowledging his obligations to Dr. R. G. Latham for several notes added to the work, and for valuable suggestions; and profits by the same opportunity to express his thanks to Captain J. E. Erskine, who has kindly placed at his disposal the MS. Journal of his Second Voyage among the Papuan Islands, affording much information in addition to the matter communicated by his "Islands of the Western Pacific," 8vo. Lond. 1853, and has added to the obligation by the contribution of several drawings of natives from his own portfolio.

1st June, 1855.

# INTRODUCTORY NOTE.

As the Editor of this work has frequently, in the additional paragraphs communicated by him, used the terms *race* and *aboriginal*, and as these words are taken by different persons in different senses, he is induced here to explain his own view of the matter. To this end he would ask leave to recite his ethnological creed, taken from what he had once intended to make a separate treatise of some extent, but which circumstances have hindered him from completing.

All mankind is the offspring of a single pair; and, to use the words of Professor Owen, "every species at every period was created most perfect in relation to the circumstances and sphere of life in which it was destined to exist." The Adamite family was what we now term civilised: it was composed of tillers of the ground, who had a settled habitat, and were guided by a systematic polity. Without some such admission, or else a perpetual interference of the Creator with the conditions of man's existence as we now find them, the human family would, in all probability, have perished. We would call this the state of nature, and all other states degraded, and would restrict the term civilised to the case where a return to that state should be effected, if this is ever the case. After the Deluge, a similar condition of things existed, slightly varied by the increased number of families who survived the great catastrophe. As the people multiplied, single families or small communities departed from the great stock, either actuated by an adventurous spirit, or driven away because of offences, or separated by accidental wanderings; and some of these detachments might continue to exist as smaller centres of population, to be left in the same way by separate bodies; and so on indefinitely. The separated communities would retain more or less of their original civil polity, as they remained together in larger or smaller masses: the extreme case of separation of single families producing mere savages,—people unable to effect anything requiring co-operation, and subsisting on the spontaneous productions of the earth, or on such animals as they could catch in the chase; while larger bodies might retain some domestic animals, and live as pastoral tribes. The original stock, remaining together, would thereby preserve their original social condition, as an agricultural people, living in settled communities.

Each one of the separated bodies is a distinct *race;* and the first race continuing to occupy any portion of the earth's surface is an *aboriginal race.* Each body received modifications from the circumstances under which it existed, and these modifications continued to accumulate as long as it continued subject to the same conditions, until the maximum of possible influence was induced; all change would be then arrested, and the race would be one homogeneous whole. This theoretical state, however, could not continue perfect, because any community, if ever so little extended locally, would be necessarily subjected to varied influences, all tending to break up the uniformity, and to give rise to what we may call sub-races; and this would take place in the original stock also, perhaps to a greater degree than with any other given race. Two communities of the same degree of separation from the original stock, exposed to similar influences, although in very different parts of the globe, would be alike in their development; and the phenomenon of like races in distant parts of the earth is perhaps more frequently attributable to this law than to immediate derivation of the one from the other. To give an instance:—tribes are found in the Asiatic islands with black skins, projecting jaws, crisp hair, and a general resemblance to the African Negro; but the two are not necessarily allied in any nearer degree than as descendants from the same original stock. In the absence of historical evidence, the language is the best test of consanguinity; there are reasons why climate should alter the physical character, but it does not appear that the language would be materially affected by such local influence.

In the early ages of man's existence upon the earth, with the world empty before him, waiting for his occupation, the process above described would be the normal condition of his increase; and the influences alluded to must have been necessarily much more powerful, and consequently the changes more rapidly brought about, than any which have been induced within the historical period. In the meantime the great original stock would grow in and round about the place where it was created, with its corresponding modifications, constituting sub-races, as alluded to above. The world would thus be gradually peopled by one dominant race, with minor variations, bearing more directly the impress of the Creator, and destined to rule over all the rest; and by many smaller races, degradations, or at least great departures, from the original type. The bodies reaching the torrid zone would be deepened in colour, and thereby better adapted to bear the heat of the sun; while such individuals as were born of lighter complexion would not long survive the exposure. Families reaching the arctic sea, finding their chief supply of food in the waters, would become fish-eaters, and would acquire the oily and obese disposition of Esquimaux and Samoiedes. A return to the original locality and to former circumstances would not imply a return to the first state: the newly induced condition would be a result of all the preceding influences. A Negro race returning to the temperate zone might lose the projection of jaw and thickness of lip characteristic of a lower social state, and produced, as physiologists have argued, by the tendencies consequent upon such a state; but there appears to be no reason why the colour should become very much modified: the race might be European in feature, and African in complexion; at all events the result would be complex, though it would be difficult to predict its exact nature.

It is obvious that migrations would bring into juxtaposition races so different that no graduation could be traced, and several races might and do exist in the same region with perfectly distinct developments. We find Hottentots, Kafirs, and Whites, at the extremity of Africa. The former may have arrived there from some early separated race, with Chinese-like features; the Kafir may have separated from a Negro stock, and lost some of the typical Negro features by residence in a more temperate climate, or may have passed through the equatorial region too rapidly to have acquired all the features which characterise his Negro brother left behind. This, of course, is merely speculative, but the general law is insisted upon. We know the history of the third race: a rapid sea-migration has not modified the Dutchman or Englishman, and the defences against local influence which the European finds or carries with him will probably preclude any very considerable change. Thus the races remain together, but distinct; and distinct they are likely to continue as long as they exist.

In this way the world would be inhabited by an indefinite number of races of men,* all graduating into each other, although, as before observed,

---

* Some ethnologists have divided the human species into three, or five, or some other number of races; but the number would really be indefinite. If the Editor were pressed by a pertinacious questioner to say how many races he would make, his reply might be, that there were as many races as persons; he might ask to be allowed to lessen by a single grade the amount of difference which his questioner should fix upon as the point of distinction, and might then apply to him the well-known verses of Horace:—

> "Utor permisso, caudæque pilos ut equinæ
> Paulatim vello, et demo unum, demo et item unum,
> Dum cadat elusus ratione ruentis acervi."

An arbitrary line must be drawn somewhere: unless this be done, if a single horsehair does not make a tail, a thousand will not. So soon as a man is born, he is potentially the head of a new race; and as all departures proceed by small steps, it must be matter of opinion at what point this is great enough to constitute a race. As convenient practically, two communities may be said to be of different races, when the change produced by circumstances is sufficiently noticeable to make the majority of individuals constituting one of them sensibly different from the majority of those of the other. It would be in accordance with general usage, also, to call two races distinct, even though they were morally and physically very much alike, if they inhabited distant parts of the world, and no connection between them were known to have ever existed.

the traces of graduation have, in many instances, been obliterated. Different races have been brought into the same neighbourhood, who have sometimes mutually destroyed each other, but perhaps oftener existed in adjoining territories, engaged in mutual hostilities, or in profitable commerce, or else amalgamating into one community under conditions now to be considered.

When two dissimilar races mix, the result is never one homogeneous race, but the characters of both are, sooner or later, distinguished in the offspring. It is stated by Dr. W. F. Edwards, that, in cases where the parents are not very unlike, the children resemble either the father or the mother, and that the features, hair, eyes, &c. of the father and mother are not blended in the children. This is true to a certain point, perhaps in the majority of cases, but very many individuals of intermediate character are also produced. It is frequently noticed in ordinary life, where the expressions "father's child" and "mother's child" are not uncommon. In the Levant, where Englishmen and Germans sometimes marry Levantine wives, a girl of dark complexion, with black hair and eyes, a decided Asiatic, will be sister of a fair flaxen-headed boy, whose every act shows his Teutonic descent; a third child of the same parents will in many cases unite the characteristics of both parents. The same intelligent ethnologist has recorded his opinion that where the parents are very dissimilar, as in the case of a White and a Negro, the union produces offspring of intermediate character, like neither parent, but combining the features of both,—a mulatto, in fact, about to become the parent of others like himself. This opposition of results, where the conditions differ in degree only, is suspicious; it gives no explanation of what is to be expected where the conditions are intermediate, where parents are not so unlike as the white and black man, nor so like as the Frenchman and German. There must be some law applicable to the whole human family, and this law the writer believes may be expressed in this way:—Considering the term *race* as before defined, "Whenever there is a union between two persons of different race, the result is ultimately a number of individuals of the two constituent races; but this result, which we will call decomposition, is more or less remote in proportion as the amalgamating races are more or less differently developed."

This law or hypothesis has been deduced from a consideration of several genealogies of half caste and mulatto families in India and America which have been communicated to the writer, and from several cases which have come under his own observation, aided by a comparison of analogous facts universally noticed by scientific sheep- and cattle-breeders. Something has been inferred from the actual observation of populations known historically to be mixed, such as our own. A large proportion of Englishmen show pure Saxon characteristics; many are equally pure Celts, or Norman

French; and very many individuals exhibit the blended features which might be expected from the continual intermarriages of such persons, some of whose children, as observed above, have the intermediate character. In the north of India, where the comparatively fair race who spoke Sanskrit are most largely grafted upon the aboriginal dark-coloured tribes, the variety of feature and complexion noticed in the same community, and even in the same family, is commented upon by observant travellers,—just as if the people, which was a mixed race a thousand years ago, were a population actually made up of individuals born in different places, and accidentally met upon the same ground.

When two individuals of remote races unite, as in India or America, the immediate offspring is almost universally of mingled characters, though, as observed by Van Amringe, who had devoted much attention to the subject, "the children [of mulattoes] seldom exhibit the medium colour of their parents." But all recorded evidence declares mulattoes or half-castes to be more liable to disease, and of shorter life, than either parent, and shows that their intermarriages are decidedly less prolific than those of other persons, though all these effects are less seen in the southern states of the American Union, and in Central America, where the white man is less white than the New Englander. Colonel Smith, in his Natural History of Man, "doubts exceedingly if a mulatto family does exist, or could exist, in any part of the tropics, continued to the fourth generation from any one stock." Dr. Knox, in his work on the Races of Men, says, "with the cessation of the supply of European blood, the mulatto of all shades must cease." There is no doubt that offspring does arise from these mixed unions, but it swerves more and more to the black and white originals, until it takes its place in one or the other body. Here, however, the writer admits the insufficiency of proof that this is owing to the working of the law of decomposition, which he has found or imagined, rather than of irregularity on the part of the wife, which is likely, and not to be judged too harshly in a class of persons so unfairly degraded and despised as those of African descent in America, or of half-blood in India. In sheep and cattle the want of permanency appears to be an admitted fact: the words of Dr. Knox are, "by no effort, saving that of constant, never-ceasing intermixture or draughts on the pure breeds, can a mixed breed be maintained;" and all practical men tell us the same.

In the case where two individuals of similar races unite—such as the Celt and Teuton,—the immediate result is, as stated before, new Celts and Teutons, with a proportion of individuals of mingled characters, in whom the decomposition is deferred to another generation, giving rise to the phenomenon so frequently seen now, and observed twenty centuries ago by Lucretius,—the likeness of a child to its grandfather or grandmother; a phenomenon which has been termed atavism.

If these statements as to the union of individuals be correct, it is obvious that when whole races are mixed, the resulting population will continue to exhibit its mixed character as long as the mixed race exists; the typical individuals will constantly produce the varieties alluded to, and the individuals of blended feature will constantly revert, sooner or later, to the original type: no mixed race of homogeneous character can be produced.

In setting down his opinion upon the result of the mixture of races, the writer has expressed himself dogmatically for the sake of brevity; but he is anxious to excite the attention of those who have leisure and opportunity for observation, rather than to enforce his own views, though his conviction of their accuracy is strong,—stronger, perhaps, than the data may appear to justify. But unless there were some principle like that which he has called decomposition, every population would be gradually reduced to uniformity, and typical forms would soon disappear altogether.

# SHORT BIOGRAPHICAL NOTICE OF THE AUTHOR,

DRAWN UP, AT THE EDITOR'S REQUEST,

BY RICHARD CULL, Esq.

OF THE ETHNOLOGICAL SOCIETY.

JAMES COWLES PRICHARD, eldest son of Thomas and Mary Prichard, was born on the 11th of February, 1786, at Ross, in Herefordshire, where his family had lived for several generations. His parents belonged to the Society of Friends.

The school education of his father, Thomas Prichard, appears to have been rather slender, for on leaving he was ill-acquainted with Latin and Greek; but, having a passion for books, he studied in private, improved his classical knowledge, acquired the French, German, and Hebrew languages, and read history. He married young, and was early left a widower with four children, to the care and education of whom he devoted much of his time.

James Cowles Prichard was never sent to school, but was educated by private tutors at home. He was taught Latin and Greek by one named Barnes; French by an emigrant, De Rosemond; and Italian and Spanish by Mordenti, a Roman. His father was most assiduous in teaching him history; and the youthful student was fond of tracing the genealogies of the kings of early times.

His father resided in Bristol for some time, for the convenience of his mercantile affairs; and it was in that city the son first manifested his inclination for those studies in which he afterwards became so distinguished. He frequented the water-side, conversed with the foreigners visiting the port, observed their physical appearance, and thus collected much information. His familiarity with French, Spanish, and modern Greek, was attributed to his habit of conversing with the foreign mariners at the water-side. The father retired from business and returned to Ross, where his son's education was continued under private tutors.

The young ethnologist chose Medicine as his profession, not from any special liking for it, but because he deemed it to be more favourable than commerce for the pursuit of that knowledge to which he was now devoted. He studied medicine at Bristol, then at St. Thomas's Hospital, London, and afterwards at the Edinburgh University, where he took his degree. Before beginning to practise his profession, he entered himself at Trinity College, Cambridge, where he appears to have read chiefly mathematics and theology, for it was while at this University that he separated himself from the Society of Friends and joined the Church of England. He, however, soon exchanged Cambridge for Oxford, and joined St. John's College, but migrated to Trinity, of which he became a gentleman-commoner.

In 1810, he settled at Bristol, in order to practise as a physician. In 1813, the first edition of his "Researches into the Physical History of Mankind," in one volume, octavo, was published. Although Dr. Prichard appears to have applied himself with zeal to the practice of his profession, as his becoming a lecturer on physiology and medicine, and his writing on medical subjects,—as Fever, Epilepsy, and Insanity,—indicate, yet his favourite study evidently absorbed much of his attention. The acquisition of the Sanskrit, Hebrew, Arabic, and Celtic languages, in a few years, proves how large a portion of his time was devoted to philology.

In 1826, the second edition of his great work was published in an enlarged form extended to double its size, in two volumes. It is impossible to compare this with the first edition without being struck with the wider and yet deeper range of his studies. In 1831 his "Eastern Origin of the Celtic Nations" was published, in which he compared their dialects with the Sanskrit, Greek, Latin, and Teutonic languages, and proved the affinity of those languages, and hence the common origin of those peoples. This was announced, in 1813, as a supplement to the first edition of his great work, but its publication was delayed, although it was then prepared for the press. As a work of original research, it is generally admitted by the most distinguished philologists to be unsurpassed in ability and soundness, while not a few deem it to be that which has made the greatest advance in comparative philology during the present century.

In 1838, Dr. Prichard published an "Analysis of the Egyptian Chronology," being a considerable extension of a former work on the same subject published in 1823. His object in entering upon the subject, "was the desire to elucidate, through the mythology of the ancient Egyptians, the relations of that people to other branches of the human family" (*Preface to the First Edition*). This was translated into German by Schlegel, who, in his preface, while praising the work for its learning and profoundness, combats the general conclusions of our author.

In 1837 appeared the first volume of the third edition of the "Researches into the Physical History of Mankind," and at successive intervals the rest of the work, which was enlarged to five octavo volumes: and during this period his "Natural History of Man" was published. He died in London after a short illness on the 22nd of December, 1848, in the 63rd year of his age, being President of the Ethnological Society at the time of his decease.

# NATURAL HISTORY OF MAN.

## BOOK I.

### General Considerations.

## CHAPTER I.

### INTRODUCTORY OBSERVATIONS.

The organised world presents no contrasts and resemblances more remarkable than those which we discover on comparing mankind with the inferior tribes. That creatures should exist so nearly approaching to each other in all the particulars of their physical structure, and yet differing so immeasurably in their endowments and capabilities, would be a fact hard to believe, if it were not manifest to our observation. The differences are every where striking: the resemblances are less obvious in the fulness of their extent, and they are never contemplated without wonder by those who, in the study of anatomy and physiology, are first made aware how near is man in his physical constitution to the brutes. In all the principles of his internal structure, in the composition and functions of his parts, man is but an animal. The lord of the earth, who contemplates the eternal order of the universe, and aspires to communion with its invisible Maker, is a being composed of the same materials, and framed on the same principles, as the creatures which he has tamed to be the servile instruments of his will, or slays for his daily food. The points of resemblance are innumerable, they extend to the most recondite arrangements of that mechanism which maintains instrumentally the physical life of the body, which brings forward its early development, and admits, after a given period, its decay, and by means of which is prepared a succession of similar beings destined to perpetuate the race. If it be inquired in what the still more remarkable difference consists, it is by no means easy to reply. By some it will be said that man, while similar in the organisation of his body to the lower tribes, is distinguished from them by the possession of an immaterial soul, a principle capable of conscious feeling, of intellect and thought. To many persons it will appear paradoxical to ascribe the endowment of a soul to the inferior tribes in the creation; yet it is difficult to discover a valid argument that limits the possession of an immaterial principle to man. The phenomena of feeling, of desire and aversion, of love and hatred, of fear and revenge, and the perception of external relations manifested in the life of brutes, imply, not only through the analogy which they display to the human faculties, but likewise from all that we can learn or conjecture of their particular nature, the superadded existence of a principle distinct from the mere mechanism of material bodies. That such a principle must exist in all beings capable of sensation, or of any thing analogous to human passions and feelings, will hardly be denied by those who perceive the force of arguments which metaphysically demonstrate the immaterial nature of the mind. There may be no rational grounds for the ancient dogma that the souls of the lower animals were imperishable like the soul of man: this is, however, a problem which we are not called upon to discuss; and we may venture to conjecture that there may be immaterial essences of divers kinds, and endowed with various attributes and capabilities. But the real nature of these unseen principles eludes our research: they are only known to us by their external manifestations. These manifestations are the various powers and capabilities, or rather the habitudes of action, which characterise the different orders of beings, diversified according to their several destinations. Among the most remarkable of these phenomena are the results of that impulse peculiar to man, which urges him to attempt and to persevere through long successive ages in the effort to obtain a conquest over the physical agencies of the elements, and to render subservient to his uses and wants the properties of surrounding bodies. While the lower tribes live every where resistless slaves to the agencies of material nature, the mere sport of their destiny, or of the lot which external conditions impose upon them, without making an effort to modify the circumstances which limit their capability of existence, man, on the contrary, gains victories over the elements, and turns the most powerful and even the most formidable of their agencies to the promotion of his own pleasure and advantage. Hence it comes to pass that man is a cosmopolite; that while, among the wild inhabitants of the forest, each tribe can exist only on a comparatively small tract of the earth's surface, man, together with those creatures which he has chosen for his immemorial companions, and has led with him in all his wanderings, is capable of living under every clime, from the shores of the Icy Sea, where the frozen soil never softens under his feet, to the burning sands of equatorial plains, where even reptiles perish from heat and drought. But here an inquiry is suggested which opens to our view a wide and interesting field of investigation. It is, whether man has not received from his Maker, besides his mental sagacity and effective contrivance, yet another principle of accommodation, by which he becomes fitted to possess and occupy the whole earth. He modifies the agencies of the elements upon himself; but do not these agencies also modify him? Have they not rendered him in his very organisation different in different regions, and under various modes of existence imposed by physical and moral conditions? How different a being is the Esquimaux, who, in his burrow amid northern ices, gorges himself with the blubber of whales, from the lean and hungry Numidian, who pursues the lion under a vertical sun! And how different, whether compared with the skin-clad and oily fisher of the icebergs, or with the naked hunter of the Sahára, are the luxurious inmates of Eastern harems, or the energetic and intellectual inhabitants of the cities of Europe! That so great differences in external conditions, by the double influence of their physical and moral agency, should have effected during a long series of ages remarkable changes in the tribes of human beings subjected to their operation, — changes which have rendered these several tribes fitted in a peculiar manner for their respective abodes,—is by no means an improbable conjecture; and it becomes something more than a conjecture, when we extend our view to the diversified breeds of those animals which men have domesticated, and have transferred with themselves from one climate to another. Considered in this point of view, it acquires, perhaps, the character of a legitimate theory, supported by adequate evidence and by an extensive series of analogous facts. But we must not omit to observe that to this opinion there is an alternative, and one which many persons prefer to maintain, namely, that the collective body of mankind is made up of different races, which have differed from each other in their physical and moral nature from the beginning of their existence. To determine which of these two opinions is the best entitled to assent, or at least to set before my readers a clear and distinct notion of the evidence that can be brought to bear upon the question, will be my principal object in the following work.

I cannot enter upon the inquiry above stated, and proceed to discuss it as a mere question of natural history, without briefly adverting, in the first place, to some considerations with which it is connected, and particularly without offering a few remarks on the relation which it bears to the history of mankind contained in the Sacred Scriptures.

# CHAPTER II.

## BEARINGS OF THE QUESTION.

THE Sacred Scriptures, whose testimony is received by all men of unclouded minds with implicit and reverential assent, declare that it pleased the Almighty Creator to make of one blood all the nations of the earth, and that all mankind are the offspring of common parents. But there are writers in the present day who maintain that this assertion does not comprehend the uncivilised inhabitants of remote regions; and that Negroes, Hottentots, Esquimaux, and Australians, are not, in fact, men in the full sense of that term, or beings endowed with like mental faculties as ourselves. Some of these writers contend that the races above mentioned, and other rude and barbarous tribes, are inferior in their original endowments to the human family which supplied Europe and Asia with inhabitants—that they are organically different, and can never be raised to an equality, in moral and intellectual powers, with the offspring of that race which displays in the highest degree all the attributes of humanity. They maintain that the ultimate lot of the ruder tribes is a state of perpetual servitude; and that, if in some instances they should continue to repel the attempts of the civilised nations to subdue them, they will at length be rooted out and exterminated in every country on the shores of which Europeans shall have set their feet. These *μιξόθηρα*, half-men, half-brutes, do not belong to what M. Bory de Saint-Vincent terms the "Race Adamique." They were made to be the domestic slaves of the lordly caste, under whose protection they are susceptible of some small improvement, comparable to that which is attained by our horses and dogs. Nothing, in the opinion of persons who maintain this doctrine, can exceed the folly manifested by the people and parliament of England when, under a mistaken impulse of what was termed philanthropy, or an erroneous notion of rights which have no existence, they committed the absurd act of emancipating from the precise condition which was most appropriate to their nature a tribe of creatures incapable of governing themselves and of combining for objects of mutual interest in a civilised community. If these opinions are not every day expressed in this country, it is because the avowal of them is restrained by a degree of odium that would be excited by it; in some other countries they are not at all disguised. Nor is it easy to prove any of the conclusions unreasonable, if only the principal fact be what it is assumed to be; if the Negro and the Australian are not our fellow-creatures and of one family with ourselves, but beings of an inferior order, and if duties towards them were not contemplated, as we may in that case presume them not to have been, in any of the positive commands on which the morality of the Christian world is founded, our relations to these tribes will appear to be not very different from those which might be imagined to subsist between us and a race of orangs. In the story of a pongo slaughtered by some voyagers in the Indian Archipelago, an account of the cries and gestures of the animal in its mortal agony, so like the expressions of human suffering, was read not without pity, and many persons censured the wanton commission of an outrage for which there appeared no adequate motive; but the capturing of such creatures with the view of making them useful slaves, even if some of them were occasionally destroyed in the attempt, would be scarcely blamed. We thus come near to an apology for the practice of kidnapping, at which our forefathers connived, though it did not occur to them to defend it on so reasonable a ground. The kind-hearted Abbé Grégoire tells us with indignation, that on the arrival of blood-hounds from Cuba in the island of Saint Domingo—"On leur livra, par manière d'essai, le premier Nègre qui se trouva sous la main." He adds, "La promptitude avec laquelle ils dévorèrent cette curée réjouit des tigres blancs à figure humaine."* Those who hold that the Negro is of a distinct species from our own, and of a different and inferior grade in the scale of organised beings, smile at the good abbé's simplicity, and observe that it cannot be much more criminal to destroy such creatures when they annoy us than to extirpate wolves or bears; nor do they strongly reprobate the conduct of some white people in our Australian colony, who are said to have shot occasionally the poor miserable savages of that country as food for their dogs.

I shall not pretend that in my own mind I regard the question now to be discussed as one of which the decision is a matter of indifference either to religion or humanity. But the strict rule of scientific scrutiny exacts, according to modern philosophers, in matters of inductive reasoning an exclusive homage. It requires that we should close our eyes against all presumptive and extrinsic evidence, and abstract our minds from all considerations not derived from the matters of fact which bear immediately on the question. The maxim we have to follow in such controversies is "fiat justitia, ruat cœlum." In fact, what is actually true it is always most desirable to know, whatever consequences may arise from its admission.

---

## CHAPTER III.

### OF GENERA, SPECIES, AND VARIETIES.

THE ancients applied the term *genus*, or γένος, to any collective number of organised beings which are akin to each other, or the offspring of the same ancestors. The idea of genus was then simple and definite, and just what we attach to the terms kind or kindred. By degrees the meaning of genus was extended, and it was made to comprehend all such creatures as by reason of some real or

* Abbé Grégoire, "Sur la Littérature des Nègres."

fancied resemblance in their form or nature were conjectured to have belonged to one original stock. Such groups were the dog-kind, the cat-kind, the ox-kind. When it was discovered, in the progress of scientific investigation, that these classes were too comprehensive, and included tribes so remote from each other that they could not with probability be regarded as the progeny of the same original tribes, the term εἶδος, or species, was adopted, and made to express nearly what that of genus had originally denoted. Species was then synonymous with stock or race. But modern naturalists have, as we shall perceive, modified the meaning of species nearly as their less scientific predecessors extended that of genus. They have attempted to found an opinion, chiefly on general resemblance, what organised creatures may or may not belong to the same tribe or kindred; and where this seemed admissible, they have termed the aggregate a species. "We unite," says M. De Candolle, "under the designation of a *species* all those individuals who mutually bear to each other so close a resemblance as to allow of our supposing that they may have proceeded originally from a single being or a single pair." "This fundamental idea is evidently built upon hypothesis."* "The degree of resemblance," he continues, "which authorises our bringing together individuals under this designation varies very much in different families; and it happens, not unfrequently, that two individuals belonging originally to the same species differ more among themselves in appearance than do others of different species. Thus the spaniel and the Danish dog are, as to their exterior, more different from each other than the dog and the wolf, and the varieties of our fruit-trees offer greater apparent differences than many distinct species of plants."

Buffon had long ago defined species in similar terms, as "a constant succession of individuals similar to and capable of reproducing each other." He here combines

* M. De Candolle, "Physiologie Végétale," tom. ii. p. 689.

two circumstances, viz. those of possible *reproduction* and of mutual *resemblance*. He had, however, previously observed that the point of *resemblance* is only an accessory idea; the single circumstance of propagation or of production from the same stock, or in other words that of supposed *kindred*, or *consanguinity*, is, in fact, the essential characteristic of species, as it originally was of genus apart from all conjectural extension of the primary meaning of that term. Cuvier adopted nearly the same definition as Buffon. He refers to mutual resemblance between individuals as a criterion of species; but species itself is fundamentally, according to both these writers, "la succession des individus qui se reproduisent et se perpétuent."*

It has been acutely observed by a writer who has of late directed his attention to inquiries connected with this subject,† that the celebrated naturalists above cited have comprehended too much in the definition of species, and, besides laying down what the term species really means, have involved an hypothetical criterion of specific identity and diversity, or of the method of ascertaining the extent and limits of these departments in organised nature.

This remark is undoubtedly well founded, not in regard only to the writers whose names have been cited, but to almost all naturalists. The adoption of a term partly of hypothetical meaning has obviously been the fruitful source of many long and intricate discussions. As the word species, apart from all hypothesis, means only what we express by kind, kindred, τὸ συγγενὲς, we might avoid a great deal of unnecessary trouble by declining the use of so disputed a term; but as we cannot banish from our vocabulary an expression so well established, we must be content to use it in its proper and restricted meaning as above pointed out.

Species, then, are simply tribes of plants or of animals which are certainly known, or may be inferred on satisfactory grounds, to have descended from the same stocks,

* Buffon, "Hist. Nat."; Cuvier, "Règne Animal."

† M. Flourens, "Annales des Sciences Naturelles."

or from parentages precisely similar, and in no way distinguished from each other. The meaning of the term species ought always, for the reasons now explained, to have been restricted to this precise import; and when the expression is used in the following pages, it is so to be understood.

The principal object of the following work may then be described as an attempt to point out the most important diversities by which mankind, or the genus man, is distinguished and separated into different races, and to determine whether these races constitute separate species or are merely varieties of one species.

Before we enter into the matter of this inquiry, it is necessary to have a clear notion of all the terms that may be used. The meaning attached to the expression permanent varieties approaches very near to that of species, and it is requisite to be careful in distinguishing the two things. Permanent varieties, it being allowed that such tribes exist, are races now displaying characteristic peculiarities which are constantly and permanently transmitted. They differ from species in this circumstance, that the peculiarities in question are not coeval with the tribe, but sprang up in it since the commencement of its existence, and constitute a deviation from its original character.

Some naturalists suppose that many of the tribes now considered as distinct species, both in the animal and vegetable kingdoms, are in reality only permanent varieties; and there seems to be little room for doubt that this opinion is in some instances well founded.

## CHAPTER IV.

### DETERMINATION OF SPECIES—PHENOMENA OF HYBRIDITY.

When any given tribes of plants or of animals are so distinguished from each other as to render it doubtful whether they belong to one species or not, several ways have been proposed for the solution of this inquiry. The most obvious and direct one is to shew that the same difference commonly, and within ordinary experience, has arisen in the same stock to which both the tribes are referred. If that can be done, the question is at once answered. But when the tribes about which the inquiry is made are either permanent varieties or separate species, there is greater difficulty in arriving at a determination.

In such instances there is one way of coming to a conclusion, which many naturalists prefer to adopt, and it is altogether satisfactory, if we can rely on the universality of an observation on which it is founded: I allude to the facts connected with what is termed hybridity.

Besides this criterion, there is another method of resolving the problem, but one which involves a long and often laborious research into the history of species. I shall have much to say on this subject after briefly surveying the phenomena of hybridity.

Nothing is more evident than the fact that all the tribes both of the animal and vegetable worlds are generally reproduced and perpetuated without becoming blended and mixed together. The law of nature decrees that creatures of every kind shall increase and multiply by propagating their own kind, and not any other. If we search the whole world, we shall probably not find one instance of an intermediate tribe produced between any two distinct species, ascertained to be such. If such a thing were discovered it would be a surprising anomaly. The existence of such a law as this in the economy of nature is almost self-evident, or at least becomes evident from the most superficial and general survey of the phenomena of the living world: for if, as some have argued, there were no such principle in operation, how could the order, and at the same time the variety, of the animal and vegetable creation be preserved? If the different races of beings were intermixed in the ordinary course of things, and hybrid races were reproduced and continued without impediment, the organised world would soon present a scene of universal confusion; its various tribes would become every where blended together, and we should at length scarcely discover any genuine and uncorrupted races. It may, indeed, be said that this confusion of all the living tribes would long ago have taken place. But how opposite from such a state of things is the real order of nature! The same uniform and regular reproduction of species still holds throughout the world; nor are the limits of each distinct species less accurately defined than they probably were some thousands of years ago. It is plain that the conservation of distinct tribes has been secured, and that universally and throughout all the different departments of the organic creation.

Strong as is the probable truth of this view of the economy of nature, it was long before naturalists were brought to admit the facts to be as they thus present themselves; and many vacillations of opinion may be traced among vegetable and animal physiologists on the subject. Among botanists, the most erroneous notions have prevailed. Linnæus, whose insight into the system of nature seemed in many respects so penetrating, adopted a very singular opinion on the extent of hybrid productions among plants. He supposed them to take place between plants of different natural families; he looked, for example, on the *veronica spuria* as the intermediate offspring of the *veronica maritima* and the *verbena officinalis;* he supposed the *saponaria hybrida* to be produced from the *s. officinalis*, fecundated by a *gentiana;* the *actea* with white fruit to be produced by the *actea* with black fruit, fecundated by the *rhus toxicodendron.* Linnæus was ready to admit facts of this kind on mere conjecture; and when he met with a plant which resembled two others that happened to grow near to it, put it down without further evidence as their hybrid offspring. These opinions have been since regarded as wholly erroneous. Attempts to produce by art such productions between plants of different families have, as M. De Candolle observes, uniformly failed, and they very rarely succeed between genera of the same family. Between species of the same genera, hybrids are, as it is well known, frequently produced in gardens; in the state of nature they are comparatively rare. M. De Candolle, after a critical examination of the examples which have been adduced, has drawn the following conclusion:—"Que, quoique l'attention des naturalistes soit éveillée depuis plus d'un siècle sur les hybrides, et que leur tendance ait paru être plutôt de les exagérer que de les réduire, on ne peut citer encore qu'une quarantaine d'exemples prouvés d'hybridité naturelle, et tous entre espèces de même genre, et même presque tous entre espèces de la même section du genre. Nous pouvons par ce fait apprécier l'hypothèse trop hardie de Linné, qui présumait que le nombre des espèces était allé en augmentant d'une manière très marquée depuis l'origine des êtres organisés, qui avait même soupçonné que le croisement des familles avait crée les genres, et que celui des genres avait crée les espèces."*

But although hybrid plants are produced, there are no hybrid races. This is a fact now universally admitted among botanists. It seems that nature has prevented the perpetuation of such productions by a variety of organic defects. M. De Candolle conjectures that the pollen of hybrid anthers is wholly or partly deficient in granules, and that on this difference depends the absolute sterility of some, and the comparative, though still defective, fecundity of other, hybrid plants. That some cause of this description must influence the results of experiments would appear evident from the observation of M. Gaertner, who found that the number of grains fertilised in each fruit is much less in the attempts to produce hybrids than in the natural process. It has been conjectured, also, by M. De Candolle, that abortion of the germs or some monstrosity in the organs of fructification, is among the causes which

* De Candolle, "Physiologie Végétale."

impede the reproduction of hybrid flowers. It appears, however, that in some instances these hybrid plants can be made to reproduce, either by blending them with the primitive kinds or with other hybrids. But this rare fertility has never been known to become permanent: according to Professor Lindley it has never exceeded the third generation. The result of all the observations which have been made upon this subject is, as M. De Candolle has remarked, that all such intermediate breeds tend incessantly to extinction, by the difficulties which are opposed to their reproduction. This explains the rarity of their appearance, and reconciles the permanence which is observed among the distinct species of nature, with the real existence, often however exaggerated, of hybrid or temporary productions, which are thus reduced into the class of monstrous and irregular phenomena in the vegetable world.*

* Mr. Knight, who has made more extensive observations on this subject than most other persons, holds most strongly the doctrine of the sterility of hybrid plants. He says, that amongst different tribes referred to the genus *Prunus*, the *Domestica*, the *Inscititia*, and *Spinosa* are likely to produce perfect offspring. He has still less doubt respecting the *Armeniaca* and *Sibirica*. The former is found in a wild state in the Oases of Africa, where it bears a rich and sweet fruit of a yellow colour; the fruit of the *Sibirica* is black, acid, and of small size. Nevertheless, he adds, "if these *apparently distinct* species will breed together, and I confidently expect they will, without giving existence to mule plants, I shall not hesitate to pronounce them of the same species, as I have done relatively to the scarlet, the pine, and the Chili strawberries." On similar grounds he infers the specific identity of the peach and the sweet almond. If the hybrid plant is productive, one of two things will be proved; either the specific identity of the two original plants, or the transmutability of the species. But if the peach were an originally distinct species, where could it have been concealed from the Creation to the reign of Claudius Cæsar? "The apple or crab of England and of Siberia, however dissimilar in habits and character, appear," says Mr. Knight, "to constitute a single species only, in which much variation has been effected by the influence of climate on successive generations." The same writer states his opinion as follows in general terms:—"I have never yet seen a hybrid plant capable of affording offspring, which has been proved, with any thing like satisfactory evidence, to have sprung from two originally distinct species; and I must therefore continue to believe that no species capable of propagating offspring, either

The history of hybrids in the vegetable creation has lately been made the subject of two comprehensive works by Gaertner and Wiegmann; and a comparative survey of the conclusions obtained by these writers, and of all that has been established in relation to the same subject, is to be found in the "Neues System der Pflanzenphysiologie" of Meyen. The following brief statement of these results is from the pen of Professor Wagner:—

"1. That hybrid plants in a natural state are very seldom produced, and that the greater number of the reputed instances rest on no sufficient evidence. 2. That hybrid plants are very seldom fruitful among themselves, but that such hybrids as the *verbascum hybridum* and the *digitalis purpurascens* from the *d. purpurea* and *lutea*, according to the corresponding observations of Koelreuter and Wiegmann, and all others which hold exactly an intermediate place between the parent plants, are absolutely barren; while those which, owing to the proportion of pollen,* partake more of either kind, and those which spring from the fertilisation of such hybrids among themselves, are occasionally propagated. 3. That plants produced from different varieties of the same species are altogether fertile, and that no impediment exists to their propagation, while hybrids either revert to the original character, generally of the maternal parent, or become gradually less capable of reproduction, and, within a few generations, entirely extinct."

A similar law prevails in the animal creation, and its effects are, on a great scale, equally constant and uniform. Mules and other hybrid animals are produced among tribes in a state of domestication; but, except in some very rare instances occurring in particular tribes of birds,

---

of plants or animals, now exists, which did not come as such immediately from the hands of the Creator,"—in other words, that no hybrid is prolific. —*Observations on Hybrids*, by T. A. Knight, p. 253 of his collected Works.

* Mr. T. A. Knight, however, was of opinion that the proportion of pollen is a matter of indifference.

they are unknown in the wild and natural state. Even when individual hybrids are produced, it is found impossible to perpetuate from them a new breed. It is only by returning towards one of the parent tribes that the offspring of these animals is capable of being continued in successive generations.

It has been shewn satisfactorily by Professor Wagner that nature has established the sterility of hybrid animals by a really organic impediment: but for the full elucidation of this subject I must refer my readers to his work on physiology.*

*Recapitulation, and application of the result.*—It seems to be the well-established result of inquiries into the various tribes of organised beings, that the perpetuation of hybrids, whether of plants or animals, so as to produce new and intermediate tribes, is impossible.

Now, unless all these observations are erroneous, or capable of some explanation that has not yet been pointed out, they lead, with the strongest force of analogical reasoning, to the conclusion that a number of different tribes, such as the various races of men, must either be incapable of intermixing their stock, and thus always fated to remain separate from each other, or, if the contrary should be the fact, that all the races to whom the remark applies are proved by it to belong to the same species.

## CHAPTER V

### OF MIXED RACES OF MEN—HISTORY OF SEVERAL MIXED HUMAN RACES.

I BELIEVE it may be asserted without the least chance of contradiction, that mankind of all races and varieties are equally capable of propagating their offspring by inter-

* A succinct statement of the facts connected with the whole of this subject has been given by Prof. Wagner in a supplemental note to his German translation of my "Researches into the Physical History of Mankind."

marriages, and that such connexions are equally prolific whether contracted between individuals of the same or of the most dissimilar varieties. If there is any difference, it is probably in favour of the latter.

If we inquire into the facts which relate to the intermixture of Negroes and Europeans, it will be impossible to doubt the tendency of the so-termed Mulattoes to increase. The Men of Colour, or the mixed race between the Creoles and the Negroes, are, in many of the West Indian islands, a rapidly increasing people, and it would be very probable that they will eventually become the permanent masters of those islands, were it not for the great numerical superiority of the genuine Negroes. In many parts of America they are very numerous, as it may be perceived by the following table, extracted from the work of M. Rugendas.*

TABLE *of the comparative Numbers of White Men, Men of Colour, Negroes, and Native Americans in different Parts of America; taken from M. Rugendas' "Voyage dans le Brézil," in folio. Paris,* 1835.

| | Year. | Whites. | Men of Colour. | Negroes. | Indians. |
|---|---|---|---|---|---|
| NORTH AMERICA. | | | | | |
| United States.... | 1820 | 7,793,008 | 1,769 | | 400,000 |
| Mexico ......... | 1824 | 1,360,000 | 2,070,000 | 8,400 | 3,430,000 |
| Guatimala ...... | 1824 | 190,000 | 320,000 | 10,000 | 965,400 |
| British Possessions. | 1822 | 1,038,000 | unknown | 5,000 | unknown |
| SOUTH AMERICA. | | | | | |
| Columbia........ | 1824 | 600,000 | 720,000 | 470,000 | 854,600 |
| Peru ........... | 1795 | 136,311 | 285,841 | 40,336 | 608,911 |
| Chili ........... | 1778 | 80,000 | unknown | 240,000 | 430,000 |
| La Plata ........ | 1824 | 475,000 including Creoles | 305,000 | 70,000 | 1,150,000 |
| Brazil .......... | | 843,000 | 628,000 | 1,987,500 | 300,000 |
| French Guiana .. | | 1,025 | 1,982 | 13,200 | 10,000 |
| British Guiana .. | | 3,421 | 3,220 | 109,349 | unknown |
| Dutch Guiana.... | | 8,525 of whom 3000 are Jews. | unknown | 72,000 | 6,200 |
| West Indies ..... | | 450,000 | 1,600,000 (Men of Colour and Negroes) | | |

* These, and some of the statements immediately following, may require modification. The Griqua Hottentots, and apparently the Cafusos, are of an origin too recent to allow of drawing any positive conclusion as to their

In order to establish this general fact of the existence of intermixed tribes descended from different races of men, I shall give a short account of several examples in which it appears that an entirely new and intermediate stock has been produced and multiplied.

The Griquas, or Griqua Hottentots, are well known to be a tribe of mixed origin, descended from the Dutch colonists of South Africa on one side, and from the aboriginal Hottentots on the other. They live on the borders of the colonial territory, and are a numerous and rapidly increasing race. The Griquas now occupy the banks of the Gariep or Orange River, for the space of at least seven hundred miles, where their numbers were estimated some years ago to be at least 5000 souls. They are powerful marauders; they harass by their predatory incursions all the native tribes in their vicinity, and are frequently troublesome to the neighbouring colonists. Great numbers of the same mixed race are in other parts thriving agriculturists; and there is a large community at Griqua Town settled under the government of the Missionaries of the United Brethren, by means of whose instructions they have been converted to Christianity and have adopted the habits of civilised society.*

The tribe of people termed by the Portuguese in the Brazils, Cafusos, are a very remarkable race, who are known to have sprung originally from a mixture of native Americans with the Negroes imported from Africa. They appear to have been accidentally separated from the other inhabitants of the country. Many families of this singular tribe now live in the solitary plains bounded by the forests of Tarama, where they were visited by the intelligent German travellers, Von Spix and Martius. From these writers we have the preceding account and the following description of the Cafusos:

---

constituting permanent homogeneous varieties, and the data regarding mulattoes are anything but definite: the transatlantic reports mix up pure negroes, if born in America, with the mixed castes.—ED.

* Thompson's "Travels in South Africa."

"Their external appearance is one of the strangest that an European can meet with. They are slender and muscular; in particular, the muscles of the breast and arms are very strong; the feet, on the contrary, in proportion weaker. Their colour is a dark copper, or copper brown. Their features, on the whole, have more of the Ethiopic than of the American race. The countenance is oval, the cheek-bones high, but not so broad as in the Indians; the nose broad and flattened, but neither turned up nor much bent; the mouth broad, with thick but equal lips, which, as well as the lower jaw, project but

FIG. 1.

Head of a Woman of the Cafusos.

little; the black eyes have a more open and freer look than in the Indians, yet are still a little oblique, and if not standing so much inward as in them, are on the other hand not turned outwards as in the Ethiopians. But what gives these Mestizos a peculiarly striking appearance is the excessively long hair of the head, which, especially at the end, is half curled, and rises almost perpendicularly from the forehead to the height of a foot or a foot and a half, thus forming a prodigious and very ugly kind of peruke. This strange head of hair, which at first sight seems more artificial than natural, and almost puts one in mind of the *plica Polonica,* is not a disease, but merely a consequence of their mixed descent, and the mean between the wool of the negro and the long stiff hair of the American. This natural peruke is often so high that the wearers must stoop low to go in and out of the usual doors of their huts; the thick hair is besides so entangled that all idea of combing it is out of the question. This conformation of the hair gives the Cafusos a resemblance to the Papuas in New Guinea; and we therefore thought it interesting to give the representation of a woman of that race in her peculiar costume."*

I shall add a description of one other race of mixed origin and strongly marked characteristics: I allude to the Papuas, spread along the northern coast of New Guinea and the adjacent islands.

The distinct aboriginal races of the Indian Archipelago and the adjoining lands may be enumerated in three divisions:—1. The Malayan or Polynesian race, whose language and physical characters are well known. They may be described as a people of tawny complexion and lank hair. This applies at least to the Malayan branch, spread over the shores of all the islands to the westward of Torres Straits. The Malays are generally supposed to have originated in the Island of Sumatra: the interior of that island is at least inhabited by people more or less allied to them, and the Menangkabau race are the genuine and probably the original Malayan stock. 2. In the interior of many of the islands, and in the mountainous parts

* Von Spix und Martius, "Reise durch Brasilien," Theil I.

of the peninsula of Malaya, it is well known that there are tribes of woolly-haired people; they are termed by the Spaniards in the Philippines, "Negritos del Monte." They have short crisp or woolly hair, and bear altogether a considerable resemblance to the Negroes of Africa; I shall term them Pelagian Negroes. 3. The interior of New Guinea, New Britain, and New Ireland, is supposed to be inhabited by a race of people who have been as yet but little known. They are termed Endamènes by the Papuas, and Alfoers, Haraforas, or Alforas, by the older voyagers.* According

FIG. 2.

to late writers they have long hair, and resemble the Australians in their physical characters, and probably belong to

* The Author was subsequently satisfied that the distinction intended by the term Harafora was not real. (See his Physical History of Mankind, Vol. v. p. 256-257.)—ED.

the same stock with that miserable and degraded race. The Papuas are distinguished from all these; they inhabit the shores of the islands of Waygiou, Sallawatty, Gammen, and Battenta, and all the northern coast of New Guinea, from Point Sabelo to Cape Dory. A singular trait in their appearance, their large bushy masses of half-woolly hair, attracted the attention of our early voyagers, and Dampier called them "mop-headed Papuas." Forrest, who describes the same people, seen by him frequently in his voyage to New Guinea, says "that the Papua Caffres are as black as the Caffres of Africa." He means the Negroes of the Mozambique coast, whom Europeans learned to term Kafirs from the Mohammedan traders in the Indian ocean. "They wear," says Forrest, "their frizzling hair so much bushed out round their heads, that its circumference measures about three feet, and when least two feet and a half." These people are clearly distinguished by Forrest from the Haraforas, and they must be equally distinct from the Pelagian Negro race, who have close hair, and are named by Dampier, in his quaint style, "shock, curl-pated New Guinea Negroes." The "mop-headed Papuas" of Dampier were first distinguished and accurately described by MM. Quoy and Gaimard, who accompanied the excellent M. de Freycinet in the expedition of the Uranie and Physicienne; and more recently by M. Lesson, who visited the same countries in the Astrolabe, and who has confirmed the opinion first advanced by Quoy and Gaimard —that the Papuas are a mixed race. These writers, says Lesson, are the first "qui ont démontré que les habitans du littoral constituoient une espèce hybride, provenant sans aucun doute des Papouas-Nègres et des Malais, qui se sont établis sur ces terres, et qui y forment à-peu-près la masse de la population. Ces Nègres-Malais ont emprunté à ces deux races les habitudes qui les distinguent. Ces insulaires forment donc une sorte de peuple métis, placés naturellement sur les frontières des îles Malaises et des terres des Papouas, et sur le littoral d'un

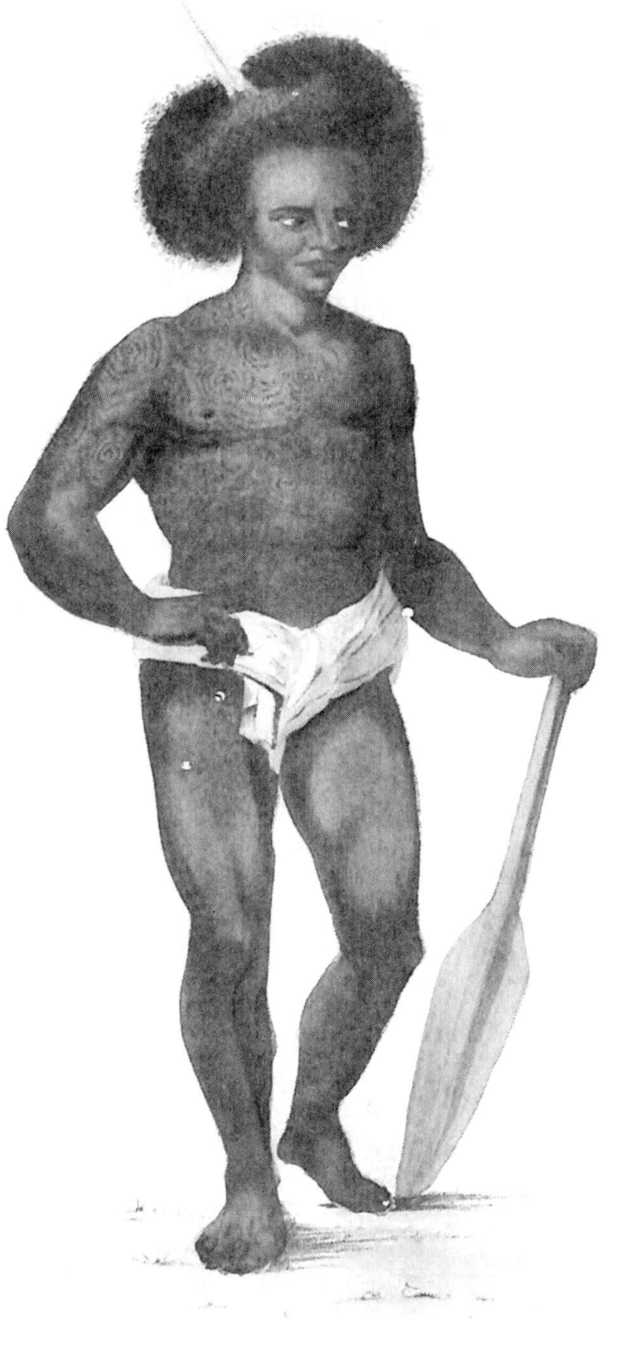

petit nombre d'îles agglomérées sous l'equateur, et au milieu desquelles s'introduisent sans interruption des Malais de Tidor et de Ternate, et des Papouas de la Nouvelle Guinee (he means the Pelagian Negroes), et même quelques Alfourous des montagnes de l'intérieur." "La masse de ces Papouas hybrides presente des hommes d'une constitution grêle et peu vigoureuse." MM. Quoy and Gaimard observe that there exists in these countries a race of people very similar to the natives of Africa, the tribes of which are interspersed among those of the Malayan race in the Archipelagos of Sunda, of Borneo, and of the Moluccas. The source of this race appears to be somewhere on the great island of New Guinea; but we must take care not to confound this race of people with that which inhabits Waygiou and the neighbouring islands, for, though these islanders resemble nearly the Negroes in the colour of their skin, they present characters which clearly distinguish them from these last. They call themselves Papuas; they have neither the hair and features of the Malays, nor those of the Negroes, but hold a middle place between both. "Leur peau est brun foncé; leurs cheveux sont noirs, tant soit peu lanugineux, très-touffus; ils frisent naturellement, ce qui donne à la tête un volume énorme, sur-tout lorsque négligeant de les relever et de les fixer en arrière, ils les laissent tomber sur le devant. Ils n'ont que peu de barbe, mêmes les vieillards; elle est de couleur noire, ainsi que les sourcils et les yeux. Quoiqu'ils aient le nez un peu épaté, les lèvres épaisses, et les pommettes larges, leur physiognomie n'est point désagréable et leur rire n'est pas grossier." The shape of the skull in this Papuan race approaches most nearly to that of the Malays, although it has some differences. Their language has never been acquired by any European; the words known appear to have no affinity with those in the vocabularies of the language spoken by the Negroes of New Guinea, as collected by the President de Brosses. *(See Plate 1, coloured.)*

It is not improbable that these tribes of the sea-coast may have come to the shores of New Guinea and the adjoining islands from some distant part of the Indian Archipelago; but whatever was the quarter whence they spread, they appear to form an example of a mixed breed of men who retain certain characters derived from their double ancestry. These traits have, however, been transmitted as permanent characteristics through many generations, since in the time of Dampier they seem to have been fully developed.

*Conclusion.*—It appears to be unquestionable that intermediate races of men exist and are propagated, and that no impediment whatever exists to the perpetuation of mankind when the most dissimilar varieties are blended together. We hence derive a conclusive proof, unless there be in the instance of human races an exception to the universally prevalent law of organised nature, that all the tribes of men are of one family.

Perhaps the solution of the problem which we have undertaken to discuss might be left on this issue, or considered as obtained by this argument. But further light may be thrown on the subject by a careful analysis of the facts which can be collected relative to the nature and origination of varieties; and it may be satisfactory to my readers to survey this field of inquiry.

## CHAPTER VI.

### GENERAL SURVEY OF THE PHENOMENA OF VARIATION IN TRIBES OF ANIMALS AND OF PLANTS.

If we could obtain a complete and satisfactory account of all the phenomena connected with the variation of breeds or races in the different tribes of organised beings, of the utmost extent to which it reaches, of its precise nature, and of the circumstances under which it takes its rise, we should experience little or no difficulty in determining the question whether the diversities which exist between different races of men are specific characters, or only examples of similar deviation. We have no hope of obtaining immediately the former of these objects to the fullest extent, but we must endeavour to approach it as nearly as possible. In proceeding in this attempt, we are first led to advert to the general fact that in the domesticated races of animals, and the cultivated tribes of plants, the phenomena of variation have been most remarkably displayed.

If we could compare our breeds of domesticated animals with the wild stocks from which they originally sprang, there would be little difficulty in ascertaining the limits to which variation in the breed can extend; but, unfortunately, opportunities for making this comparison are difficult to obtain, and in some instances altogether beyond our reach. The original stocks of our domesticated animals are rarely to be recognised in their primitive state among the wild animals of the earth; we know not what has become of them, unless it be supposed that they have been wholly subdued by man. There are, indeed, wild oxen, sheep, goats, horses; but the most of these are tribes which appear to have returned in some degree to their original state, after having been more or less completely domesticated. We are ignorant of the time and circumstances under which most of these races became wild, and of the particular breeds from which they descended. There is, however, one great field of observation in the tribes of animals which are known to have been transported from Europe to America since the discovery of the western continent by the Spaniards in the fifteenth century. Many of these races have multiplied exceedingly on a soil and under a climate congenial to their nature. Several of them have run wild in the vast forests of America, and have lost all the most obvious appearances of domestication. The wild tribes are found to differ physically from the domesticated breeds from which they are known to have issued, and there is good reason to regard this change as a restoration in part of the primitive characteristics of the wild stocks from which the tamed animals originally descended.* The comparison of these wild races with our domesticated breeds affords at least some curious and interesting observations. The animals which were transported by the Spaniards to America are the hog, the horse, the ass, the sheep, the goat, the cow, the dog, the cat, and gallinaceous fowls. On each of these we shall collect some observations relative to the changes which it has undergone in becoming restored to the wild state.

Hogs, as we learn from M. Roulin, were introduced into St. Domingo at the first discovery of that island by Columbus, in November 1493, and successively to all the places where the Spaniards formed settlements. The first which appeared on the high plains, or table-land, of Bogota, came by a very indirect route: they were not carried up the river Madelaine by Quesada, but were actually brought from Peru by the soldiers of Benalcazar, one of Pizarro's adventurous followers, who, while wandering a whole year in search of the fabulous El Dorado, took with them male and female pigs to stock their future colony. These animals multiplied so rapidly that, in the space of half a century, they were spread from the twenty-fifth degree of north latitude to the fortieth degree south. In St. Domingo, herds of swine so overran the country that, on

* Of these naturalised tribes in America, as well as of the indigenous races, some brief accounts were given long since by Don Felix de Azara, in his well-known work on Paraguay; but it is to a recent work of M. Roulin, published in the "Mémoires de l'Institut," among the "Contributions des Savans Etrangers," that we are indebted for more exact and valuable information.

M. Roulin spent six years in Columbia, during which time he was partly occupied in collecting the information communicated in his memoir. His works are professedly confined to New Grenada and Venezuela, countries traversed by the Cordillera of the Andes, and containing within them great diversities of climate. For all our information respecting the wild tribes in Paraguay we are indebted to Azara.

the introduction of the sugar-cane, it was found necessary to destroy them in great numbers. In less than thirty years after the discovery of America, we learn from Oviedo that herds of wild swine infested the woods of Cuba, Porto-Rico, and Jamaica. That writer supposed them not to exist on the continent, owing to the wild beasts which destroyed them; but M. Roulin has seen wild hogs in the great plains of the Llanos, spread far to the eastward of the Cordillera, particularly on the left bank of the river Meta, a country where couguars and jaguars are very numerous.

These animals, wandering at large in the vast forests of the New World, and feeding on wild fruits, have resumed the manner of existence which belonged to the original stock: their appearance nearly resembles that of the wild boar. Their ears have become erect; their heads are larger, and the foreheads vaulted at the upper part; their colour has lost the variety found in the domestic breeds; the wild hogs of the American forests are uniformly black. The hog which inhabits the high mountains of Paramos bears a striking resemblance to the wild boar of France. His skin is covered with thick fur, often somewhat crisp, beneath which is found in some individuals a species of wool. From excessive cold and defect of nourishment, the hog of that region is of small and stunted figure. In some warm parts of America the swine are not uniformly black, as above described, but red, like the young pecari. At Melgara and other places there are some which are not entirely black, but have a white band under the belly reaching up the back: they are termed *cinchados*.

The restoration of the original character of the wild boar in a race descended from domesticated swine removes all reason for doubt, if any had really existed, as to the identity of the stock; and we may safely proceed to compare the physical characters of these races as varieties which have arisen in one species (*see figs.* 3 *and* 4). The restoration of one uniform black colour, and the change of thin, sparse hair and bristles for a thick fur with a covering of wool, are facts

FIG. 3.

Wild Boar.

that must be noticed in the observations of M. Roulin. The difference in the shape of the head between the wild

FIG. 4.

Hampshire Sow.

and domestic hog of America is very remarkable. Blumenbach long ago pointed out the great difference between the cranium of our swine and that of the primitive wild boar; he remarked that this difference is quite equal to that which has been observed between the skull of the Negro and the European. "Those persons," he says, "who have no opportunity of verifying the fact, have only need to cast their eyes on the figure which Daubenton has given of both the former (*see figs.* 5 *and* 6). I shall pass

Fig. 5.

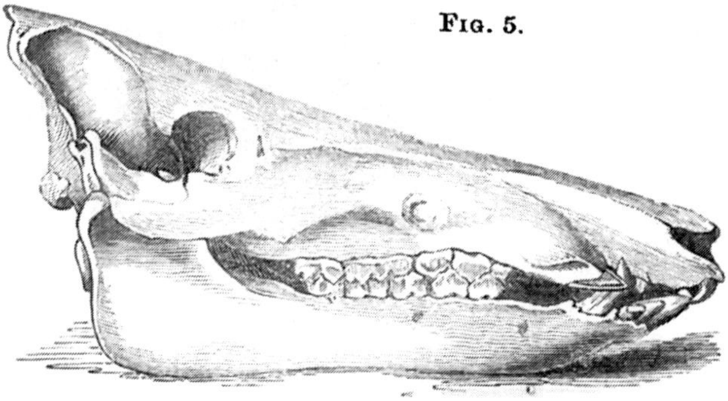

Skull of a Wild Boar.

over," he adds, "the lesser varieties of breed which may be found among swine, as among men, and only mention that I have been assured by M. Solzer that the

Fig. 6.

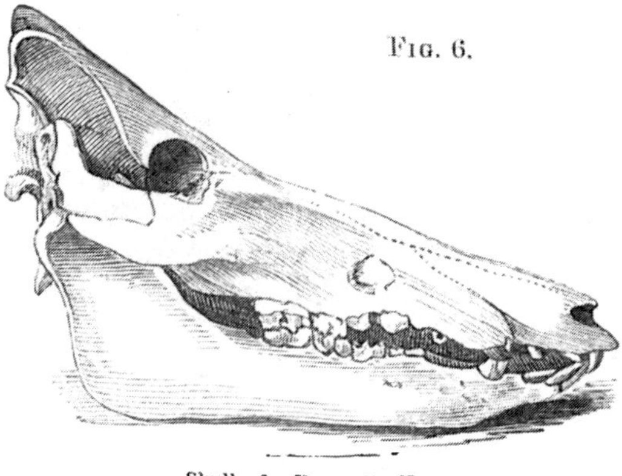

Skull of a Domestic Hog.

peculiarity of having the bone of the leg remarkably long, which in the human kind is observed among the Hindús, has been remarked with regard to swine in Normandy.

They stand very long on their hind legs; their back, therefore, is highest at the rump, forming a kind of inclined plane; and the head proceeds in the same direction, so that the snout is not far from the ground."

"Swine," continues Blumenbach, "in some countries have degenerated into races which, in singularity, far exceed every thing that has been found strange in bodily variety among the human race. Swine with solid hoofs were known to the ancients, and large breeds of them are found in Hungary and Sweden. In like manner the European swine, first carried by the Spaniards in 1509 to the Island of Cubagua, at that time celebrated for its pearl-fishery, degenerated into a monstrous race, with toes which were half a span in length."

There are breeds of solid-hoofed swine in some parts of England. The hoof of the swine is also found divided into five clefts.

Buffon had before remarked the varieties of the hog tribe: "In Guinea," he observes, "this species has acquired very long ears, couched upon the back; in China, a large pendant belly, and very short legs; at Cape Verde and other places, very large tusks, crooked like the horns of oxen; in domestication, half-pendant and white ears."

Horned cattle were introduced into St. Domingo in the second voyage of Columbus. In twenty-seven years there were, according to Oviedo, herds of 8000 head; in 1530 they were slaughtered merely for their hides, of which, if we may believe Acosta, the annual exportation amounted, in 1587, to more than 35,300. In the same year upwards of 60,000 hides were exported from New Spain.

It was long ago observed by Don Felix de Azara that the wild oxen of South America differ in colour from the domestic breeds in the same country. The herds of tame cattle, he says, display great varieties of hue, but the colour of the wild oxen is constant and invariable; the upper parts are of a brown-red, and the rest of the body black: he adds, that he hence suspects the primitive colour of the species to have been what the Spaniards term "*hosco*." He notices an instance of the origination of a particular breed in this country without horns. "En 1770 il naquit un taureau *mocho*, ou sans cornes, dont la race s'est très multipliée."* When the bull has no horns, the calves are also destitute of them.

In some of the hot provinces of South America, M. Roulin informs us that a variety of ox has been noted for an extremely rare and fine fur. These oxen are termed "pelones." The variety is reproduced or descends in the stock, but is not cultivated, because the pelones are too delicate in constitution to bear the cold of the Cordillera, to which the cattle are driven for the provision of the towns there situated. The pelones evidently constitute a variety adapted to a particular climate. Oxen of other breeds often perish when driven into the same provinces, and are with difficulty assimilated to the climate, or *acclimatised*. In the same hot countries a variety is sometimes produced with an entirely naked skin, like that of the dogs without hair found at Calongo, or Cacongo, on the coast of Guinea. These cattle are called Calongos: they are very delicate and weak. This variety never makes its appearance in cold districts.

A very remarkable fact relative to the oxen of South America is recorded by M. Roulin, to which M. Geoffroy St.-Hilaire has particularly adverted in the report made by him on M. Roulin's memoir before the Royal Academy of Sciences."† In Europe, the milking of cows is continued through the whole period, from the time when they begin to bear calves till they cease to breed. This secretion of milk has become a constant function in the animal economy of the tribe; it has been rendered such by the practice, continued through a long series of generations, of con-

* "Voyages dans l'Amérique Méridionale," par Don F. de Azara, tom. i. p. 378.

† "Annales du Muséum," tom. xxvii.

tinuing to draw milk long after the period when it would be wanted for the calf: the teats of the cow are larger than in proportion, and the secretion is perpetual. In Columbia, the practice of milking cows was laid aside, owing to the great extension of farms and other circumstances. "In a few generations," says M. Roulin, "the natural structure of parts, and withal the natural state of the function, has been restored. The secretion of milk in the cows of this country is only an occasional phenomenon, and contemporary with the actual presence of the calf. If the calf dies the milk ceases to flow, and it is only by keeping him with his dam by day that an opportunity of obtaining milk from cows by night can be found." This testimony is important by the proof which it affords that the permanent production of milk in the European breeds of cows is a modified function of the animal economy, produced by an artificial habit continued through several generations.

Two other very important observations made by M. Roulin in South America were pointed out by M. Geoffroy St.-Hilaire in his report to the Academy of Sciences. They refer to the fact of the hereditary transmission of habits originally impressed with care and art upon the ancestors. Of this fact I shall adduce other examples in the sequel; at present I only advert to M. Roulin's observations. The horses bred in the grazing farms on the table-land of the Cordillera are carefully taught a peculiar pace, which is a sort of running amble. This is not their natural mode of progression, but they are inured to it very early, and the greatest pains are taken to prevent them from moving in any other gait; in this way the acquired habit becomes a second nature. It happens occasionally that such horses, becoming lame, are no longer fit for use: it is then customary to let them loose, if they happen to be well-grown stallions, into the pasture grounds. It is constantly observed that these horses become the sires of a race to which the ambling pace is natural, and requires no teaching. The fact is so well known, that such colts have received a particular name: they are termed "*aguilillas*." The second fact is the development of a new instinct, which, as M. Roulin declares, seems to become hereditary in the breed of dogs found among the borderers on the river Magdalena, which are employed in hunting the pecari. I shall cite the author's own words :—" L'adresse du chien consiste à modérer son ardeur, à ne s'attacher à aucun animal en particulier, mais à tenir toute la troupe en échec. Or, parmi ces chiens, on en voit maintenant qui, la première fois qu'on les amène au bois, savent déjà comment attaquer; un chien d'une autre espèce se lance tout d'abord, est environné, et quelle que soit sa force, il est dévoré dans un instant."*

It appears that barking is an acquired hereditary instinct; it has become natural to domesticated dogs and young whelps to learn to bark even when separated at birth from their parents. It has been conjectured that barking originated in an attempt to imitate the human voice; however that may be, wild dogs do not bark. There are numerous troops of wild dogs in South America, principally in the Pampas. There are also in the Antilles, and in the isles on the coast of Chili, similar breeds. These breeds, in recovering their liberty, have lost the habit of barking; like other uncultivated breeds of dogs, they only howl. It is known that the two dogs brought to England by Mackenzie from the western parts of America could never bark, and continued to utter their habitual howl; but a whelp bred from them in Europe learned to bark. It has often been observed that the dogs in the island of Juan Fernandez, the progeny of those which were left there purposely by the Spaniards before Lord Anson's time, with the design of exterminating the goats, were never known to bark. A curious observation of M. Roulin is, that the cats in South America have, in like manner, lost those "*miaulemens incommodes*" which are so often heard during the hours of night in many parts of Europe.

* P. 339.

The ass has not run wild in any part of South America visited by M. Roulin, but wild herds of horses exist in many districts of Columbia, and in the high plains near the source of the Rio Negro. In some high districts, though the pasturage is abundant, the race diminishes in size; they are covered with a hair or fur so thick as to render them hideous. Azara informs us that the wild horses of Paraguay are all of one hue, while the tame breeds are in that country, as elsewhere, of various colours. The wild horses are all of a chestnut or bay-brown colour. "Cela pourrait faire penser que le cheval original ou primitif était bay-brun."*

The breed of sheep that was transported into America by the Spaniards was not the Merino, but that termed the sheep "de lana burda y basta."† There are numerous flocks on the Cordillera between 1000 and 2500 mètres of elevation. The hot country between the river Meta and the foot of the Cordillera is uncongenial to the breed of sheep. In the valley which separates the eastern chain of the Cordillera from the middle chain, the flocks are not numerous; but the sheep of this region display a phenomenon which is worthy of observation: wool grows on the young lambs nearly as in temperate climates; if shorn, it sprouts again and the fleece is formed as usual; if neglected, it forms itself into a large tufted mass, which breaks off in shaggy portions. When it comes off there is found beneath, not fresh wool, nor a naked and diseased skin, but a short fine hair, shining and smooth, like that of the goat in his best state, and this remains permanent, the wool never reappearing.

The goat in South America has become more agile, of more slender make, with the head better formed, and bearing smaller horns than in Europe. The most marked sign of domesticity in our European goats, viz. the large size of the teats, has completely disappeared in the South American goats.

* Azara, *ubi supra*, p. 374.

† With harsh and coarse wool.

We are informed by Azara that sheep and goats bear twice in the year in South America, and produce at least three lambs or kids annually.

The introduction of geese into America affords an instance of the process of acclimatisation. About twenty years since geese were first introduced on the plateau of Bogota. At first the eggs laid were very few, and scarcely a fourth part were hatched; of the young goslings more than half died in the first month; the second generation produced by the survivors was more successful, and the breed gradually approximates to the vigour of the same stock in Europe.*

A similar remark has been made respecting gallinaceous fowls. According to Garcilasso, in the valley of Cusco they had been often introduced many years before pullets could be hatched from the eggs laid. This stock has now become quite productive, but English game-fowls lately introduced produce but two or three pullets at most in a brood. These two breeds display another difference; the long acclimatised breed, or, as M. Roulin terms them, the "creole pullets," whose ancestors have lived for ages in a hot climate, have chicken nearly naked, or without down, and this is the case till their feathers grow. The newly imported English breed is covered with close down. "Le petit animal est encore vêtu comme pour vivre dans le pays d'où ses pères ont été apportés depuis peu d'années."

The fowls of Nicaragua are a breed well known at Bogota, in which, according to our author, a strong predisposition appears to what he terms "melanism," namely a production of the black colour in the young brood.

* A parallel observation was communicated to me by Mr. Rankin, who became acquainted with the fact at Sierra Leone. English wheat sown there brings forth stalks of which the ears contain very few grains; these again sown give origin, in the second generation, to a more productive growth. The wheat becomes acclimatised in a few generations in intertropical Africa, after undergoing a similar process to that which the geese of Europe sustain before the breed becomes adapted to the climate of South America.

These fowls are termed Negro pullets. As their colour is disadvantageous, there is no temptation to propagate them, yet they are very common. It would seem, as he observes, that besides the individuals who inherit this colour, there is a strong tendency to originate it in the ordinary breed. In connexion with this remark, M. Roulin has made a general observation, which is important; it is, that throughout intertropical America, both *melanism* and *albinism*, as he terms the black and white varieties, make their appearance very frequently in warm-blooded animals, and that these two descriptions of "*monstrosity*" are among the peculiarities which are most readily transmitted to the offspring. "Perhaps this remark," he adds, "is equally applicable, in the fullest extent, to the region which is situated at the antipodes of that described. It holds, at least, as far as it relates to black pullets, and we are informed by Marsden, that in Java the gallinaceous fowls are often affected with melanism, and many travellers assure us that albinism often appears in the human species in the Sunda isles." He might have alluded to the white elephants of Siam. According to M. Roulin, the melanism of the fowls of Bogota is less remarkable in the skin than in the crest and periosteum, the serous membranes, and the cellular web which surrounds the muscles. Similar observations have been made respecting the black fowls of Malabar. The facts adduced by this writer appear to him to afford sufficient evidence to establish the following general conclusions:—

1. That when some animals are transported into a new region, not only individuals, but races, require to be harmonised in physical constitution to the climate.

2. This acclimatisation, as it is termed, consists in certain permanent changes produced in the constitution of animals, which bring it into a state of adaptation to the climate.

3. A restoration of domestic animals to the wild state causes a return towards the original characters of the wild tribe.

Two other remarks may be added, as incidental results from the facts reported by M. Roulin :—

1. Permanent changes or modifications in the functions of animal life may be effected by long-continued changes in the habitudes which influence these functions. This is proved by the fact relative to the milking of cows.

2. Hereditary instincts may be formed; some animals transmitting to their offspring acquired habits, and the psychical as well as the physical characters of races undergoing variation through the agency of various causes on the breed. All such variations are within a limited range.

## CHAPTER VII.

### PHENOMENA OF VARIATION CONTINUED—VARIOUS BREEDS OF DOMESTIC ANIMALS IN THE OLD CONTINENT.

THE series of facts which we have surveyed, relating to changes developed in breeds of animals transported to the New World within the space of three centuries, affords, perhaps, the best authenticated examples of the effects which a change of external conditions is capable of producing in races of animals subjected to its influence. They serve, at least, to show of what nature are the deviations which we may expect to discover under similar circumstances. If a longer period of time had been given, it may be supposed that greater diversities would have displayed themselves. Of this we shall be convinced on comparing the various breeds of the most anciently domesticated animals of the Old Continent. We cannot, indeed, point out in every instance the same positive proof of the descent of these several breeds from a common original, but there is sufficient evidence to justify our assuming this fact, in several cases, as one extremely probable. The differences observable between the breeds of domestic animals are very considerable, if we compare the extreme instances; but they are also very numerous, and display so many intermediate links, that there is no strongly marked line of separation, such as we generally find on comparing distinct species: they pass into each other by almost imperceptible degrees. Moreover, the extent of diversity in structure, or of the development of organic variety, and of the corresponding improvement in the instinct and other animal faculties, is generally in proportion to the progress of domestication and of the degrees in which the cultivated state of the tribe differs from its wild and original condition, to the care which has been bestowed in bringing it to its improved state, and to the length of time that may have passed since it was reclaimed. We have, indeed, in many cases, unquestionable proofs that such changes in organisation and habit have, in fact, resulted from alterations in the manner of life, and especially from a removal into a new climate; the effects of such deviations on the breed having been notoriously the obliteration of particular characters and the development of others.

I shall illustrate these remarks by adducing some additional facts displaying the influence of external causes on breeds of animals, and by describing briefly the most remarkable instances of variety observed in the several domesticated species.

### *Varieties in the Breeds of Sheep.*

The sheep is one of the most anciently domesticated animals, and it is one in which great varieties display themselves. It has been long believed, and this appears to have been the opinion of Baron Cuvier, that all the breeds of tamed sheep are descended either from the argali of Siberia, or from the mouflon or musmon of Barbary. This is, at present, doubted by most naturalists. There seems, however, to be no reason for believing that the domestic breeds belong to more than one species, though they differ much in different countries. In Europe the breeds of sheep vary much in stature, in the texture of their wool, the number and shape of their horns, which are in some large, in some small, in others wanting to the female, or altogether absent from the breed. The most important varieties in Europe are the Spanish breeds, some with fine, others with crisp wool, in which the rams have long spiral horns (see *fig.* 7); the English breeds, which

FIG. 7.

Spanish Sheep.

differ greatly in size and in the quality of the wool; and in the southern parts of Russia, the long-tailed breed. The breeds of sheep in India and in Africa are remarkable for the length of their legs, a very convex forehead, and pendant ears; these also have long tails; their covering is not wool, but a smooth hair. In the northern parts of Europe and Asia the sheep have short tails. The breeds spread through Persia, Tartary (see *fig.* 9), and China, have their tails transformed into a double spherical mass of fat. The sheep of Syria and Barbary (see *fig.* 8), on the other hand, have long tails, but likewise loaded with a mass of fat. In both of these varieties of the sheep the ears are pendant, the horns of the rams large, and those of the ewes and lambs of moderate size, and the body is covered with wool mixed more or less with hair.*

FIG. 8.

Fat-tailed Sheep of Syria.

It has been asserted from time to time that particular breeds of sheep retain their peculiarities when transported into various climates differing from those where the breed prevails, but the assertion is not true to the fullest extent; it admits, at least, of exceptions. Pallas, who described the sheep of the Kirghis Tartars, which have a remarkable structure and display the type of the Tartar breed in an exaggerated degree, added a particular intimation that these animals retain their peculiar characters in whatever climate they may be placed after being removed from their native pastures, which are on some of the highest parts of Central Asia. He says that sheep are not elsewhere found so large and deformed in appearance as these of the Kirghis. They are taller than a calf at its birth, and very heavy, and in their proportions they bear some resemblance to the Indian breeds. Their heads are very protuberant, and their ears large and pendant; their lower lips project much beyond the upper. Most of them have warty excrescences covered with hair hanging from their necks.

* Cuvier, "Règne Animal," tom. i.

Instead of a proper tail, they have a large round mass of fat almost without hair, beneath.* The intelligent German traveller M. Ermann, during his late journeys in Northern Asia, had an opportunity of correcting an error in the information given to Professor Pallas. He assures us that the fat-tailed sheep of the Kirghis, when transferred into Siberia, do not preserve their peculiarities. The dry and bitter herbage of the Steppes is unfavourable to the growth of fat, and they lose the mass of adipose matter. This was observed near Schaitansk, to the northward of Tscharensberg. Even in the Southern Ural in the pastures of Orenburg these sheep lose their fat tails *after a few generations.*† (See *fig.* 9.)

FIG. 9.

Fat-rumped Sheep of Tartary.

New breeds of sheep are frequently formed in different countries in which particular qualities predominate, according to the preference of the breeders. This is done partly by crossing or intermixing races already constituted and well known, but in great part also by selecting individuals from the stock in which the particular qualities are more

* Pallas, "Reise durch Siberien," &c. French translation.

† "Reise um die Erde," von H. Ermann.

strongly marked than in the generality of the same breed. In these instances the natural or congenital variety, which the individual animal displays perhaps for the first time, becomes perpetuated by the hereditary transmission of such characters, which is a law of the animal economy. A striking instance of this fact is to be found in the origination of a new breed of sheep in the state of Massachusetts, which has been noticed by many writers in connection with this subject. In the year 1791, one ewe on the farm of Seth Wright gave birth to a male lamb, which, without any known cause, had a longer body and shorter legs than the rest of the breed. The joints are said to have been longer, and the fore legs crooked. The shape of this animal rendering it unable to leap over fences, it was determined to propagate its peculiarities, and the experiment proved successful; a new race of sheep was produced, which, from the form of the body, has been termed the otter-breed. It seems to be uniformly the fact, that when both parents are of the otter-breed, the lambs that are produced inherit the peculiar form.*

*Horses.*—Horses are found wild in some parts of Asia and of Africa, but it is difficult to ascertain in such instances whether they have always remained in their original state, or whether the troops which wander in the forests, alienated from human care, may have descended from individuals of a domesticated breed which have escaped at some unknown time. It is much disputed what was the primitive country of the species. The breeds of domesticated horses have great varieties of form in different parts of the world, but they all partake so entirely of all the manifest and particular characters which are accounted specific, and the different breeds pass so into each other by imperceptible gradations, that no doubt is entertained in general of their

* Only one case has been reported as an exception to this remark, and that was questionable. For an account of the facts relating to the otter-breed of sheep, I refer the reader to a memoir by Colonel Humphries, "Philosoph. Transact." 1813; and to Thomson's "Annals of Philosophy."

all belonging to one species. They differ, however, greatly in stature if we take the extremes, and likewise in other peculiarities. The largest breeds are chiefly found in northern countries, in Tartary and the north of Europe; those of Arabia and of Barbary are of light and agile form: the breeds in various countries of Europe are formed from different intermixtures of these, and by the careful propagation of particular qualities which manifest themselves from time to time, and which are of the nature of accidental or natural varieties.

Wild horses are well known to have proportions of body somewhat different from the most improved races. Their heads are larger, and the foreheads are of a round and arched form; their hair is rough, long, and crisp.* It has been observed by Blumenbach, that there is less difference in the form of the skull in the most dissimilar families of mankind than between the elongated head of the Neapolitan horse and the skull of the Hungarian breed, which is remarkable for its shortness and the extent of the lower jaw. In this country the heads, as well as the whole form of the skeleton of race-horses, differ much from those of the draught-horses. It was remarked by Pennant, that wild horses have heads larger in proportion, and foreheads of a round and arched form. Pallas has confirmed and illustrated this observation by an account of a race descended from horses which have run wild in Eastern Siberia in the vast plains near the sources of the Tschugan. These animals, which are the remote offspring of domesticated horses, now differ from the Russian breed in having larger heads and more pointed ears; their mane is likewise short and bristly, and their tail has become shortened. Their colour is said to be almost uniformly of a dun or brown; pied and black horses are very rarely seen among them. Pallas adds, that their principal traits, or those which distinguish them from domestic breeds of the horse-kind, and which may be considered as characters

* Pennant, "History of Quadrupeds;" Cuvier, "Règne Animal," tom. i.

acquired by the race since it ran wild in the desert, are the following:—They have larger heads than domestic horses, with more vaulted foreheads; their mouths are more hairy, and the mane comes down lower on the shoulders; their limbs are stronger; their back less arched and straighter; their hoofs are smaller and more pointed; their ears are longer and are bent more forward.*

*Oxen.*—The breeds of the ox-kind are very numerous and diversified. The wild original appears to have been lost, for the urus, or aurochs, has been proved by Cuvier to be a different species. The fossil skulls, which are supposed to be the relics of ancient wild animals of the ox species, have the horns bent forward and downward. But the size and direction of the horns vary much in the domesticated breeds, and this character is taken as the distinguishing trait of many races; as of the middle-horned, and the long and short-horned breeds in England, of which the

FIG. 10.

Lancashire Ox.—Specimen of the Long-horned Breeds.

former are supposed to be an ancient British race, and descended from the stock of the black cattle of Wales and

* Pallas, "Reise," *ubi supra.*

Scotland. (See *figure* 10). In Abyssinia, the breeds of oxen are remarkable for the enormous size of their horns. In Paraguay, Azara remarked, with surprise, that the breed of oxen, though descended from the horned race, are destitute of horns,—a circumstance which he contrasts with the fact, much more extraordinary, if true, that horses are sometimes seen in the same country bearing horns.

Considerable diversities in the shape of the head and the proportion of parts have been described by Sturm, Meckel, and others, in the different breeds of European oxen.*

*Dogs.*—"The domestic dog," says Baron Cuvier, "is the most complete, the most singular, and the most useful conquest that man has gained in the animal world: the whole species has become our property; each individual belongs entirely to his master, acquires his disposition, knows and defends his property, and remains attached to him until death; and all this, not through constraint or necessity, but purely by the influence of gratitude and real attachment. The swiftness, the strength, and the sharp scent of the dog have rendered him a powerful ally to man against the lower tribes, and were, perhaps, necessary for the establishment of the dominion of mankind over the animal creation. The dog is the only animal which has followed man over the whole earth."

Some naturalists have supposed the dog to be of the same species with the wolf; others, to be a domesticated jackal: both these tribes more nearly resemble the dog than does the fox. It is yet doubted by many whether the dog and the wolf are distinct races, though M. de Serres has pointed out characteristics which he considers as specific differences between the dog and his three congeners.† But are all the different breeds of dogs varieties

* J. F. Meckel, "Traité d'Anatomie Comparée." Sturm, "Racenzeichen der verschiedenen Hausthiere," &c.

† The orbit, in particular, is larger in the dog than in the fox, and in the fox than in the wolf. See M. Marcel de Serres, Memoir on the Distinctive Characters of the Dog, the Wolf, and the Fox, as suggested by the Skeleton.—*Ed. Phil. Trans.* July 1835.

of one stock? This has often been questioned. The best naturalists, and those who have devoted the most attention to the history of the species, are of this opinion. M. Frédéric Cuvier has strongly defended it; he observes that if we assume the varieties to be permanent races, or originally distinct species, and predetermine that these races are susceptible of few or no modifications, it will be requisite to lay down the existence of at least fifty separate species of dogs, all distinguished from each other by permanent characters.

Few persons, says M. Cuvier, can seriously give credit to so improbable a supposition; it becomes, in fact, more difficult of admission when we advert to the series of progressive changes observable on comparing the physical conformation of the different breeds of dogs. The least domesticated races, and those which have become wild, as the Dingo or Australian dog, differ little in the shape of their skulls and in other characters from the wolf; while the more cultivated breeds, or those which have their faculties most developed and their habits most changed by domestication, deviate in the same proportion from this form, and, in particular, exhibit a much more vaulted and arched forehead, and a greater development of the brain. The Australian dogs are almost in the natural and wild state; they live in holes of rocks, and support themselves, independently of man, by catching wild prey; and when they hunt in company with the native Australians, it is rather as associates in the chase, from which they are rewarded with a share of the booty, then as trained and domesticated animals.* The muzzle of the Australian dog

* "Bien différent de nos chiens domestiques, il n'a aucune idée de la propriété de l'homme. Il se jette avec fureur sur la volaille, et semble ne s'être jamais reposé que sur lui-même du soin de se nourrir." "Ne nous offrent-ils pas le tableau que Buffon peint de l'homme et du chien sauvage s'entr'aidant pour la première fois, poursuivant de concert la proie qui doit les nourrir, et la partageant ensemble après l'avoir atteinte?"—F. CUVIER, *Sur le Chien des Habitans de la Nouvelle Hollande*, &c. "Ann. du Muséum," tom. xi.

is not shortened like that of the bull-dog, nor elongated like that of the greyhound, but resembles the muzzle of

FIG. 11.

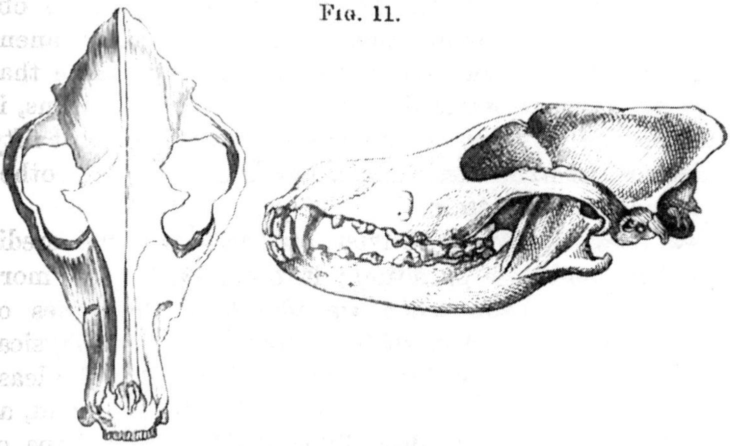

Skull of an Australian Dog.

the mastiff; their ears are erect, moveable, and having the opening in the anterior part; the senses of smell and hear-

FIG. 12.

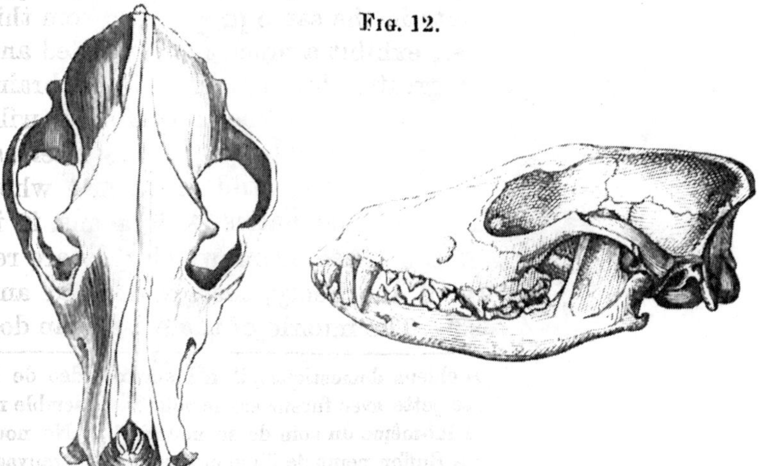

Skull of a Mastiff.

ing are proportionably acute. They are gregarious, and sometimes hunt in troops of 200 in a body, and admit not the approach of dogs which do not belong to their society.

The skull of the Australian dog differs but little from that of a wolf; in both the head is very flat, and the cavity which contains the brain has comparatively very little space. This arises from the flattening of the temporal and parietal bones, which from their outer and lower margins pass almost in a level plane towards the median line, where they join the opposite bones with very little elevation, thus forming a flattened roof for the cerebral cavity. This may be seen in the accompanying sketch. The Danish dog and the mastiff resemble in the shape of their heads the Australian, and they display as little development of intellect or sagacity. (See *fig*. 12.) The terrier and the hound differ from the preceding breeds in having the parietal bones much more arched, and allowing a larger space for the brain. (See the following figures.) The greyhound has a larger muzzle and smaller frontal sinuses than the hound; and the sense of smell is proportionally deficient in this breed.

FIG. 13.

Scottish Terrier.

The shepherd's dog, which displays much greater sagacity than the hunting dogs above mentioned, and which Buffon very erroneously considered as the least modified by domestication, has a very considerable capacity of the cranium.

The temporal bone in the head of the shepherd's dog is not flattened from the inferior margin, nor rounded off with a

FIG. 14.

Shepherd's Dog.

trifling degree of arching or elevation towards the opposite side; in the shepherd's dog, the bones rise perpendicu-

FIG. 15.

Greyhound.

larly to one-half of their vertical extent, and then become arched over the space occupied by the brain. The wolf dog resembles the shepherd's dog. Again, in the spaniel and water-dog, the capacity of the cranium is much greater

FIG. 16.

Old English Hound.

than in the shepherd's dog; and these races, in all their varieties, are remarkable for a great development of the

FIG. 17.

Wolf Dog.

frontal sinus, which is so considerable as to give the outline of the forehead a direction almost perpendicular to that of the nasal bones: the lower jaw is very much bent. The head of the bull-dog differs remarkably from that of all

FIG. 18.

Newfoundland Dog.

the preceding varieties; the posterior parts of the system of facial bones are situated higher than the muzzle, and the

FIG. 19.

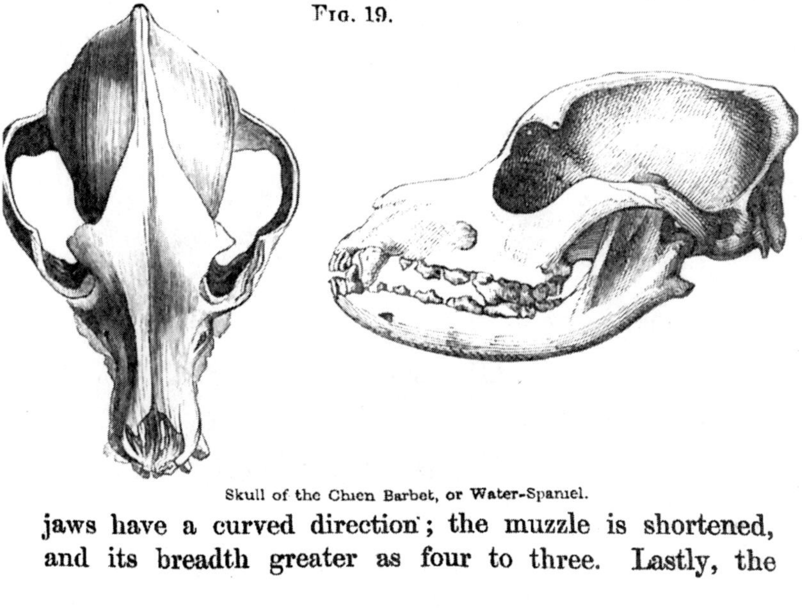

Skull of the Chien Barbet, or Water-Spaniel.

jaws have a curved direction; the muzzle is shortened, and its breadth greater as four to three. Lastly, the cranium of the bull-dog is much less capacious than that of the shepherd's dog; and the parietal bones, instead of being

FIG. 20.

Bull-Dogs.

arched, bend towards each other almost at right angles. The docility of these races evidently bears a due proportion

FIG. 21.

Iceland Dogs.

to the capacity of their skulls. The wolf-dog, and the spaniel and water-dog, display wonderful intelligence, and seem to understand the voice of men. Restored to a state of comparative wildness, which approaches to their unreclaimed and primitive condition, the tribes of dogs everywhere make a corresponding approximation to the type which may be supposed to have belonged to the species in its original state.

The several varieties of dogs differ much in stature, or size, in the shape of their ears or their tails, which have from sixteen to twenty or twenty-one vertebræ, varying in particular breeds. Some tribes have an additional toe or claw in the hind foot, as some human families have six fingers; and many dogs have an additional or false molar tooth, either on one or the other side. The

FIG. 22.

Dalmatian Dog.

hair, or the covering of the skin, varies greatly in different tribes of dogs; the colour, the fineness of texture, the length and position, are all different. Dogs of cold climates have generally two kinds of hair,—a fine woolly hair, close to the skin, and long silky hair; in tropical climates, the former lessens, and finally disappears altogether; and the same thing happens in our dwellings which afford shelter from inclement temperatures. The Turkish dog, or more properly the dog of Guinea, has a naked and oily skin; the bull-dog, the greyhound, and hound, have short and smooth hair; the shepherd's dog, the Australian dog, the mastiff, and the Iceland dog, have longer hair than the above, but much shorter than the spaniel, water-dog, and lap-dog; some breeds of dogs, again, have woolly and tufted hair: in short, the dog-kind presents all the varieties in respect to the nature of the hairy covering of the body that are to be found in the entire class of mammiferous animals.* It must be observed that these varieties in the hair, as well as other traits in the breed of dogs, have in the first instance a relation to climate, but have yet the character of permanent varieties, which remain for generations constant and undeviating, like the corresponding peculiarities in different races of men; for we find nearly all the varieties of dogs propagated in the same climate without any very notable change, when the breeds are left distinct; the physical as well as the psychical qualities of each stock being transmitted with little variation to its posterity. The varieties of the dog tribe have become *permanent varieties*.

It was long ago observed by Pallas, that no domesticated animals exhibit greater variety than gallinaceous fowls. "Some breeds," he says, "are large, some extremely small; they are tall, dwarfish; have small or large and double combs; some have tufts of feathers on their head; some have bare and yellow legs, others have their legs covered with feathers. What is still more remarkable is, that there is a breed without rumps, common in some parts of England, and another with five claws." The fowl of Padua, of which Pallas has published an account, has a peculiarity in the conformation and capacity of the skull, which is, perhaps, a greater deviation from the usual structure than any other species of animals presents.†

* "Recherches sur les Caractères Ostéologiques qui distinguent les principales Races du Chien domestique." Par M. Frédéric Cuvier. "Ann. du Muséum," tom. xviii.

† Pallas, "Specilegia Zoologica," 4to.

# CHAPTER VIII.

THEORY OF VARIATION IN THE SPECIES OF ANIMALS AND OF PLANTS — NISUS FORMATIVUS — DIFFERENT MANIFESTATIONS OF THIS TENDENCY. 1. VARIETIES IN ORGANIC STRUCTURE. 2. IN THE FUNCTIONS OF THE ANIMAL ECONOMY. 3. IN PSYCHOLOGICAL CHARACTERS.

After surveying the most obvious phenomena of variety in races, we must attempt to draw some inferences respecting the nature of these deviations, and to lay down a few general observations as to their extent.

It is certain that we must regard such varieties not as fortuitous or accidental phenomena, but as the results of a particular tendency, or of a process in the animal economy productive of changes which are necessary for the continued existence of tribes or races under certain external conditions. Blumenbach, who first observed this tendency, gave it the term of "Bildungstrieb," or "Nisus formativus," and the latter designation has been adopted by M. Geoffroy St.-Hilaire, and other writers on natural history. It is a vital power, or the result of vital arrangements in living bodies, in virtue of which organisation receives a peculiar direction from external circumstances, and sometimes exhibits very remarkable deviations from its general uniformity. As an exemplification of this new formative or organising process, Blumenbach refers to the production of galls in various trees, as the rose-gall. The insertion of the rudiment of an insect into the bark of the rose-tree, in this instance, gives rise to that curiously organised growth termed the rose-gall, which is so unlike anything produced by the regular and undisturbed economy of vegetable organisation, and so similar to all other rose-galls, and to the growth produced by similar causes on other trees. Such growths, however apparently anomalous, are all regular, and follow a definite line. Trees of various species have their particular galls. Organisation here receives a new direction under particular circumstances; so does it when plants or animals are by any means brought under conditions different from those under which they may have originally and usually existed.

The facts which I have already collected plainly indicate that similar changes have, for the most part, a tendency to preserve each tribe, or to maintain its continued existence under new conditions. Such is the theory of acclimatisation; but the change of climate is only one of the occasions on which this tendency is called into operation. It is likewise displayed in the change of habits induced by domestication. Domestication seems to effect a much greater change in the whole manner of existence than any removal from one country to another that can be imagined to take place during the continuance of the wild state. Its results are, in matter of fact, far more extensive on the nature of animals. Domestication is not a casual and temporary change effected in an individual, but the modification of a race, by which it becomes fitted to exist under new circumstances. It has often been observed, that under domestication the natural instincts of the wild tribe appear to be either obliterated or greatly altered. It has been well remarked by Dr. Hancock, that dogs appear to have acquired, through long association with man, some traits of character which can be considered in no other light than as imitations, or perhaps as feeble gleams or scintillations, of reason.* These phenomena, as that excellent writer has shown, are quite different in kind from the psychical properties which belong to any tribe of animals whatever in the wild and natural state. Such properties, and the characteristics of domestication in general, cannot be produced in any one animal caught from the woods or the mountains, and tamed as far as one individual is capable of being tamed. The wide difference in habits and instincts which we perceive on comparing the domestic dog with the original, or rather from the nearest approximation

* Dr. Hancock on Instinct.

to the original type that can be discovered, or of which we can form an idea by studying the history of the Australian and other wild tribes of the species, can only have been the sum or result of a series of changes carried on through many generations, every successive generation displaying a character more and more in harmony or in relation with the conditions under which the tribe is brought to exist. The final cause of such a provision in nature is obvious, since without it domesticated animals, properly so termed, could have had no existence. But we do not refer to the theory of final causes as establishing the fact, though there is no difficulty in connecting such a probable argument with the principles of the inductive philosophy. The fact rests on the evidence of phenomena actually observed.

We shall obtain a further insight into the nature of the changes produced by acclimatisation and domestication, and shall have an opportunity of adding some further proofs in aid of these conclusions, if we distribute the whole of the phenomena of variation under three different heads.

1. Differences of organic structure, including all the varieties of external appearance, colour, form, stature, and proportion of parts, which are perceptible in creatures belonging to the same original stock. 2. Physiological differences, or varieties in bodily temperament, internal constitution, and the function of the animal economy. It may seem at first unlikely that such variations as these should take place in the offspring of one species; but if we consider the differences which actually exist even between individuals of the same family, born and bred under the same circumstances, it becomes no longer improbable that greater deviations should occur under different conditions. 3. Psychological varieties, or diversities in psychical properties, that is, in instincts, habits, intellectual and moral characters, in so far as such attributes can be predicated of the inferior animals.

### 1. *Varieties in Organic Structure.*

Varieties in form and structure are observable in the offspring of the same parents; and as there is through all organised tribes a very perceptible tendency to the permanent transmission of any bodily peculiarities that once display themselves, we have thus a foundation laid for varieties of breeds: in fact, there are no tribes of animals existing without such varieties. But these deviations from a common type are all within certain limits, and leave permanent and unaltered the specific character of the tribe. It is not always an easy matter to determine what the specific character is, and what properties are variable; in general, those characters are most permanent which have the greatest influence on the habits and psychological character of the tribe, as the number and form of limbs, or organs of motion and progression, the organs of sense, the number and disposition of the teeth. External characters, such as colour, the properties of the integument, stature, and the proportionate length and size of limbs, are more liable to deviation from the common standard. It is well known that these varieties are most numerous and remarkable in tribes which have been domesticated, and have continued to exist and to propagate their stock under circumstances often very remote from those which are natural to the race in its wild and unrestrained condition. All the species of animals which have been found susceptible of domestication are consequently divided into a great number of various breeds, while among the untamed and untameable inhabitants of the desert there is comparatively very little diversity. The dog, which has been the companion of man from the earliest times, and has followed him into all climates, displays, perhaps, the most strongly marked and numerous varieties. Between the different breeds of dogs we discover the greatest variations, both in structure and in psychological characters; and in both these respects the dog has been contrasted with elephants, which have seldom been propagated in captivity, but are caught afresh from the wilderness, and, as we might expect, display very little deviation from the original and common type of organisation.

The shape of the head furnishes, in the structure of the bones, the most remarkable instances of variety, and some of the leading characters which distinguish particular races. This has been noticed by many comparative anatomists, and particularly by J. F. Meckel, and by Sturm, who has written a work professedly devoted to the varieties of form in the breeds of oxen. The proportional length and thickness of the neck are likewise characteristic of particular races, more especially in the breeds of horses. Meckel likewise observes that the length, height, and proportional breadth of the hinder parts serve to distinguish different tribes from each other, as do also the length and thickness of the tail. The pelvis is in proportion broader or narrower in particular breeds, and this is transmitted to posterity, and becomes a constant character. Varieties also occur in the comparative length of the limbs, and in the proportions which they bear to each other and to the whole body.

In particular systems, or textures, or parts of the body, varieties are also found. These are regarded by Meckel as of inferior importance, but they are observed to be frequently transmitted, and to become permanent characters. Of this kind are varieties in the texture and development of the epidermis and the other parts of the common tegument. Scales, feathers, hairs, and in part horns, may be considered as subordinate to this texture, to which the epithelium corresponds, in the internal surfaces.

The disposition, form, and development of those parts of the system on which are dependent the processes connected with reproduction and suckling, undergo similar variations. Under this head M. Meckel alludes to the elongation of the mammary organs, and to the depositions of fat, which are remarkable in some races both of human beings and of cattle in South Africa.

The stature and mass of the body in general also characterise particular races, as in the different breeds of horses, oxen, sheep, and dogs. The last especially displays, in these respects, singular varieties.

Colour, especially that of the skin and its appendages, presents distinctive characters, though, perhaps, more variable and partial ones. One hue, though in different shades, is often common to a whole race.*

### 2. *Physiological Varieties or Diversities in the Internal Constitution.*

Individuals differ so much in these particulars, that we may easily conceive differences to exist between races long separated, though sprung from the same original stock. A certain uniformity of constitution, or a constant adherence, within a particular range of variety, to certain laws of the animal economy, belongs to the specific character of each original race. Particular species have certain limits with regard to the average duration of life, the circumstances connected with reproduction, such as the number of their progeny, the times and frequency of breeding, the period of gestation in mammifers, and in birds that of sitting upon eggs, and in the length of time during which they suckle or watch over their young. The progress of physical development and decay is likewise ordained by Nature to take place in each species according to a certain rule. The periods at which individuals arrive at adult growth, the different changes which the constitution undergoes at particular ages, the periods of greatest vigour and of decline, and the total duration of life, are given, though with individual exceptions and varieties, to every species of animals. There are exceptions and varia-

* "Traité Générale d'Anatomie Comparée." Par J. F. Meckel. Traduct. Paris, 1828.

tions, but these are within certain prescribed limits, and obey definite laws.

On the other hand, it may be observed as a very general fact, that animals belonging to tribes which nearly resemble each other, but are yet specifically distinct, differ in a decided manner with respect to the same particulars.

From what has been said, we must expect to find varieties even in these particulars within the limits of one species, and yet characteristic of tribes. The observation, before cited from M. Roulin, as to the difference between the cows of South America and those of Europe in the time of giving milk, may prepare us for finding similar deviations.

A number of facts might be adduced from the history of endemic or local diseases, proving that tribes of men who have remained for many generations in some particular region have acquired a constitution different from that of the first settlers. Disorders break out among them to which the original colonists were not liable. The capability of contracting such complaints does not exist in the race until after an abode continued through several generations in the district where they are endemic; but at length the race becomes entirely acclimatised, and equally liable with the other inhabitants to the diseases with which the latter have been long familiar.*

### 3. *Psychological Characteristics.*

The instincts and habitudes of animals have been much studied as particular subjects of curiosity and speculation, but not so much as characteristic of species. In the inferior tribes these phenomena are most wonderfully diversified:

* Cretinism, prevalent in the Valais, and in the valley of the Lena in Eastern Siberia, does not attack families which have not been resident for some generations in the infected districts. A parallel observation may be made as to many other constitutional disorders, as plica Polonica, frambæsia or yaws, and some species of lepra or elephantiasis. For particulars and proofs of these remarks, I must refer the the reader to my "Researches into the Physical History of Mankind," vol. i.

the habits of particular tribes are known to be peculiar, and different from those of other races of the same group, but scarcely anything has been observed as to variety in psychical characters within the limits of a particular species. These are two different series of facts which deserve to be considered. It seems likely that in the state of nature these properties, like those of external character, are uniform, but that they also are susceptible of variation through the interference of man.

Among the insect tribes, it is most remarkable that the several species display peculiar habitudes and various modes of energy, extending to everything within their sphere of action and of existence: they have different methods of providing for their young; construct their nests of different materials; place them in different situations; deposit their eggs, and protect them in different ways: each peculiarity in the habits of the species being common to all the individuals comprised in it. In order to be fully convinced of this remark, we have only to read the admirable descriptions given by Messrs. Kirby and Spence, of the *hymenoptera*, and particularly of the wild bees and wasps. The *xylocopa violacea*, which bores cylindrical tunnels in the trunks of trees; the *melitta fodiens*, which perforates earthy banks; the *apis manicata*, which places her eggs in membranaceous coverings in holes; the *apis muraria*, which builds for them stone walls; the *apis papaveris*, which covers them with the leaves of the wild poppy; the *centuncularis*, or *Rosenbiene*,* which lines her burrows with rose-leaves, are all different species of bees, distinguished from each other by their specific habits, more strongly than by any discovered peculiarity in the structure of body belonging to each tribe. Analogous varieties of instinct distinguish the different species of wasps, among which the *odynerus muraria* is remarkable, as well as the several species of *cynips*, or *Gallwespe*,† of which one tribe produces the rose-galls, another

* Blumenbach, "Handbuch der Naturgeschichte."

† Ibid.

those of the oak, and a third the galls of the *carica*, or wild fig; and likewise the several races of *tinea* and *curculio*. Each species in these several tribes obeys laws or principles of action entirely peculiar to its own kind, and distinct from those which govern all other kinds. Other families of insects and arachnoids are equally diversified by habits appropriate to their respective species; almost every species of spider is distinguished by a peculiar method of spinning its web.

Among the most surprising instincts of mammiferous tribes are the migratory propensities of the lemmings, or wandering rats. The migrations of these tribes are well known to be executed with surprising energy, and with the universal accord of the whole tribe. But to each tribe the migratory movement, as well as other habits, is peculiar. The lemmings of the Scandinavian Alps are not found far towards the East; even in Russian Lapland they are unknown; and they are replaced near the shores of the Polar Seas, and in the Ural, by a race different in aspect and in colour, and smaller by at least one-third. These races may be considered as species nearly approximating; they are distinguished by a striking difference of instinct. The Scandinavian lemmings are said to lay up no provisions, and to have only a single chamber in their dwelling-places; while those of the Uralian tribe excavate numerous apartments, and provide for the winter season by storing up magazines of the rein-deer lichen.*

There are instances of psychological characters appropriated to particular species among the tribes of animals with which we are more familiarly acquainted. Nothing is more remarkable in the dog-kind than the inclination of the whole tribe to associate themselves with man; whence it has arisen that in all ages, and in almost every corner of the globe, they have been his companions and devoted slaves. In this respect the dog is strikingly contrasted

* "Dictionnaire Class. d'Hist. Nat." art. Campagnol. "Quarterly Review," vol. xlvii. p. 339.

with his congeners, the wolf, the fox, and the jackal. The fierce and untameable disposition of the wolf separates him at a wide distance from the dog, and his gregarious habits equally distinguish him from the solitary fox. The psychological distinctions are in these instances, perhaps, more strongly marked than are those of anatomical structure. Even in the instance of sheep and goats, which are generally reckoned as distinct genera, it has been observed that psychical characters constitute the most striking difference. The sheep, always stupid, or of the most simple understanding, from its birth timid and inert, follows its dam the same feeble and defenceless animal that it is destined to remain through life,—an emblem everywhere of unresisting innocence. The goat, agile and ever roaming, shows its disposition not less early: the young kid, driven by its instinct, seeks in the first hours of its existence the clefts and summits of rocks which Nature already points out to it as its future abode.

It would appear that every species of animals has a definite psychological character, which is at least as appropriate and typical of the particular race as are any characters whatever of organisation. The psychological character, as far as it is connected with organisation, is, indeed, the highest and ultimate result of the arrangements of structure belonging to each living body, and may therefore be expected to be distinctive and typical of it. But the type is preserved with individual varieties, as we easily perceive in every domesticated species, and the uniformity of psychical character appropriate to each tribe is susceptible of certain shades of variation. These last are chiefly noted in species diversified by the effects of domestication, varieties of character following for the most part varieties of organisation, as already noted in the dog-tribe. But the instances in which we have the best opportunity of observing the origination of these characters are those in which they may be termed artificial instincts. Some examples of this kind have been cited from M. Roulin in the breeds of horses and of dogs in South America. A very clear and authentic testimony, establishing parallel facts, is afforded by Mr. T. A. Knight in some of the memoirs which detail his experiments and observation on the breeding of animals. Mr. Knight says, "The offspring of domesticated animals inherits in a very remarkable manner the acquired habits of their parents. In all animals," he adds, "this is observable, but in the dog it exists to a wonderful extent; the offspring seems to inherit not only the passions and propensities, but even the resentments of the family from which it springs. I ascertained by repeated experiments, that a terrier, whose parents had been in the habit of fighting with polecats, will instantly show every mark of anger when he first perceives the scent of that animal, though the animal itself be wholly concealed from his sight. A young spaniel brought up with the terriers showed no marks whatever of emotion at the scent of the polecat, but it pursued a woodcock the first time it saw one with clamour and exultation; and a young pointer, which I am certain had never seen a partridge, stood trembling with anxiety, its eyes fixed, and its muscles rigid, when conducted into the middle of a covey of those birds. Yet each of these dogs is a mere variety of the same species; and to that species none of the said habits are given by nature. The peculiarities of character can therefore be traced to no other source than the acquired habits of the parents, which are inherited by the offspring, and become what I shall call instinctive hereditary propensities. These propensities, or modifications of the natural instinctive powers of animals, are capable of endless variation and change; and hence their habits have become adapted to different countries and different states of domestication, the acquired habits of the parents being transferred hereditarily to the offspring. Bees, like other animals, are probably susceptible of these changes of habit, and thence, when accustomed through many generations to the hive, in a country which does not afford hollow trees or other habitations adapted to their purpose, they may become more dependent upon man, and rely on his care wholly for a habitation; but in situations where the cavities of trees present to them the means of providing for themselves, I have found that they will discover such trees in the closest recesses of the woods, and at an extraordinary distance from their hives, and that they will keep possession of such cavities in the manner I have stated."

The preceding observations were detailed in a paper read before the Royal Society in 1807; and in 1837, just thirty years afterwards, the author again addressed the Society on the same subject, and in confirmation of his previous remarks. He then declared that he had commenced his experiments on dogs sixty years before, and that they occupied a good deal of his attention during twenty years, and to a less extent nearly to the date of this memoir. "In a communication," he says, "which I had the honour to address to this Society upon the economy of bees, I gave an opinion that the families of those insects, in common with every species of domesticated animal, are, to a greater or less extent, governed by a power which I have there called an 'instinctive hereditary propensity;' that is, by an *irresistible propensity* to do that which their predecessors of the same family have been taught or constrained to do through many successive generations." "I had at that period made a great many experiments analogous to those which were then related, and I have subsequently made a considerable number. As no person is now likely to investigate this subject so laboriously, or through so long a period, I believe that the facts which I am prepared to communicate may deserve to be recorded in the Transactions of this Society.

"At the period in which my experiments commenced, well-bred and well-taught springing spaniels were abundant, and I readily obtained as many as were wanted; but within a short time some facts very strongly arrested my attention. In several instances, young and wholly inexperienced dogs appeared very nearly as expert in finding woodcocks as their experienced parents. Woodcocks are driven in frosty weather to seek their food in rills of unfrozen water. I found that my old dogs knew as well as myself what degree of frost would suffice for this effect, and as this knowledge proved troublesome, I left them at home, and took only the wholly inexperienced young dogs; but I found to my astonishment, that they confined themselves to the unfrozen ground, just as their educated parents would have done. I was led to conclude that the young dogs were governed by feelings and propensities similar to those of their parents."

Mr. Knight in this memoir, to which I must for want of room for extracts refer my readers, has mentioned several instances in which dogs appeared to have derived extraordinary powers from the highly cultivated instinct, or, as he terms it, intellect of their parents.

He has detailed analogous facts respecting some other animals: "The hereditary propensities of the offspring of Norwegian ponies, whether full or half-bred, are very singular. Their ancestors have been in the habit of *obeying the voice of their riders*, and not the bridle, and the horse-breakers complain that it is impossible to produce this last habit in the young colts; they are, notwithstanding, exceedingly docile and obedient when they understand the commands of their master. It is equally difficult to keep them within hedges, owing, perhaps, to the unrestrained liberty to which the race may have been accustomed in Norway."

Mr. Knight devoted much attention, as I have observed, to the economy of bees. In these he has shown that the natural instinct of bees drives them to make their nests in hollow trees; but that they will abandon such a receptacle if a hive is offered to them. This, however, he concludes, is "rather habit, produced by domestication through many successive generations, than anything inherent in the nature of bees, which induces them to accept a hive when offered them in preference to the situation which they had previously chosen." He has noticed the disposition to migrate to exist in a much greater degree in some families of bees than in others.

An equally remarkable effect of domestication, though a more ordinary one, and on that account alone less striking, is the change of temper, if I may be allowed the expression, which the whole breed undergoes. Perhaps this, if duly considered, affords as strong a proof as can be sought, of an hereditary modification of psychical character. For the tameness of domestic animals is not to be attributed to the early teaching which they receive, or the state of subjection in which the young are reared; their congenital dispositions must be altered. The cub of a wild boar, taken from its dam at the hour of its birth, would be in disposition very unlike a sucking-pig. A similar difference has been observed in the young of wild and tame rabbits, though this species is one which is very little altered in form by domestication. A person in the habit of breeding animals assured me that he has taken young rabbits from burrows soon after their birth, and has bred them up in confinement, feeding them with a spoon. Yet the young of the wild rabbit, though similar in colour and form, were very unlike the brood of the domesticated; they were not tame, though bred up in captivity.

## CHAPTER IX.

### CONCLUDING OBSERVATIONS ON THE THEORY OF THE VARIATION OR DEGENERATION OF ANIMALS.

FROM the preceding survey of the phenomena of variation in the tribes of animals, and of the circumstances under which these appearances are displayed, we may venture to draw the following general inferences.

1. That tribes of animals which have been domesticated by man and carried into regions where the climates are

different from those of their native abode, undergo, partly from the agency of climate, and in part from the change of external circumstances connected with the state of domesticity, great variations.

2. That these variations extend to considerable modifications in external properties, colour, the nature of the integument, and of its covering, whether hair or wool; the structure of limbs, and the proportional size of parts; that they likewise involve certain physiological changes or variations as to the laws of the animal economy; and lastly, certain psychological alterations or changes in the instincts, habits, and powers of perception and intellect.

3. That these last changes are in some cases brought about by training, and that the progeny acquires an aptitude to certain habits which the parents have been taught; that psychical characters, such as new instincts, are developed in breeds by cultivation.

4. That these varieties are sometimes permanently fixed in the breed so long as it remains unmixed.

5. That all such variations are possible only to a limited extent, and always with the preservation of a particular type, which is that of the species. Each species has a definite or definable character, comprising certain undeviating phenomena of external structure, and likewise constant and unchangeable characteristics in the laws of its animal economy and in its psychological nature. It is only within these limits that deviations are produced by external circumstances.

Races of men are subjected more than almost any race of animals to the varied agencies of climate. Civilisation produces even greater changes in their condition than does domestication in the inferior tribes; we may therefore expect to find fully as great diversities in the races of men as in any of the domesticated breeds. The influence of the mind must be more extensive and powerful in its operations upon human beings than upon brutes; and this difference transcends all analogy or comparison. *A priori*, we might expect to discover in the psychological characters of human races changes similar in kind, but infinitely greater in degree.

In the following chapters of this work I shall proceed to survey the phenomena of diversity displayed by human races in all the three points of view to which I have adverted. In each respect I shall ascertain whether there is a common specific type preserved amidst all the varieties which display themselves, and shall endeavour to determine whether the differences which are found when remote and diversified human races are compared with each other, are such as fall within the limits of the principle of variation.

---

## CHAPTER X.

### OF THE DIVERSITIES OF ORGANISATION IN DIFFERENT RACES OF MEN; AND, FIRST, OF VARIETIES IN THE COMPLEXION AND IN THE STRUCTURE OF THE SKIN.

We now proceed, according to the arrangement above mentioned, to consider the varieties of organisation which distinguish the races of men, — differences of structure coming first in order, and preceding those of the physical functions and the phenomena which refer to the higher principles of life. We here find, in the first place, two classes of facts most striking to our observation: first, the diversities in external parts, in the colour and texture, or in the qualities of the superficial or outer covering of the body; secondly, varieties in the internal parts, in the proportions which they bear to each other, and the construction of the bony fabric. In order to form a correct idea of these diversities, and to estimate their bearing on the general question to be discussed, we shall find it necessary to survey them collectively in the several aggregates in which they are displayed, as the characters of particular races. But there are some points connected with this subject which require a more detailed examination than this method allows. We must, in the first instance, consider accurately the intimate nature of those organic peculiarities on which the variety of complexion, and of the qualities of the integument and of the hair, depends; pointing out, as we proceed, circumstances which illustrate the relation of these phenomena to the question of diversity or sameness of species. We shall afterwards advert to the varieties of form and configuration, and divide human races, chiefly with reference to this part of the subject, into several groups.

The varieties in the colour of the body, and the texture of the inner and outer integuments, depend on the organisation of parts which are in one sense extra-cutaneous. These parts belong to what is sometimes termed the horny covering of the body, and are often said, though erroneously, to be in their nature inorganic, or at least not possessed of proper vitality; they are really endowed with a peculiar vital condition, and with a very remarkable and curious mode of organisation, the leading characters of which have lately been ascertained by microscopical researches, though much, perhaps, remains to be done before they can be fully and satisfactorily elucidated.

### *Of Varieties in Colour or Complexion.*

It has generally been thought that the differences of colour or of complexion are less important in the discrimination of races than other characters, particularly than varieties in the form of the body and in the shape of the skull. A French writer of the present time, M. Flourens, well known for the extent and accuracy of his researches in various subjects connected with anatomy and physiology, considers the difference of colour as more essentially characteristic of distinct races than any other peculiarity. The grounds of this opinion will be made apparent to the reader in the following pages.

It is a common remark, that the complexion or the

colour of the skin has a certain relation or correspondence to that of the hair and to the hue of the eye, or rather of the iris; this is true as a general observation; but there are many exceptions to the remark, and particularly as it respects individuals and races who have black hair. Among Europeans the two most strongly marked varieties of complexion are those which the French term "blonde" and "brunette;" one having blue eyes, light flaxen hair, and a fair skin; the other, black eyes, a browner skin, and black hair. As we know no expressions in English precisely correspondent with these names, blonde and brunette, I have adopted those of xanthous and melanous as distinguishing terms. To these two varieties we must add a third, the leucous or the albino, which is regarded as a kind of monstrous or defective phenomenon, being much more rare than the preceding. The greater number of persons in the middle countries of Europe are neither properly blondes nor brunettes, xanthous nor melanous, but have a complexion which is between these two extremes. In many northern parts the xanthous predominates, and in the south the melanous. If we divide human races into these three varieties, founded chiefly on the colour of the hair, we must consider the melanous as comprising great diversities in the colour of the iris and in the shade of the skin. In various nations who have generally black hair, the iris is often of a deep brown or chocolate colour, as among the Chinese; in many it is frequently of a green or hazel hue, as in some tribes of Negroes in Kongo; in some black-haired people it is grey and even blue. All these are variations from the more prevalent hue, which is black when the hair is of that colour; the variations which tend towards the development of a light complexion sometimes affecting only the eyes, while the skin remains of a very dark hue; in other examples the skin is also white, or, as it were, etiolated; in some the hair itself varies, and becomes yellow or red, and this even in black-skinned races: the hue of the skin, however, generally assumes, under the same circumstances, a lighter shade. These variations occur, as we shall prove by examples, in the offspring of melanous, and even of black races. But similar changes make their appearance even in individuals. Children born fair, and continuing to have light brown hair during youth, often become black-haired as they approach the adult age. A similar transition has been known to take place from the leucous to the xanthous complexion; that is, albinos cease to be such. The hue of the eye in the leucous, or albino, is red: the black pigment which lines the choroid, as well as the colouring substance of the iris, being defective, a red tinge is imparted to the light which penètrates the transparent blood-vessels of the iris and the interior parts of the eye. This defect, joined to a total want of colouring matter in the hair and the skin, constitutes a true albino. When the colouring matter, which was during infancy defective, is produced in after-life, the xanthous complexion takes the place of the leucous, as, in more frequent examples, the melanous supersedes the xanthous.

The following observations, illustrative of this remark, are extracted from an excellent paper by Professor Graves, of Dublin:—"Last year Dr. Ascherson informed me that he had seen a case of the after-development of the pigment of the eye in an albino boy three years old. This child had at its birth white hair and violet-coloured eyes with dark red pupils; at the end of the third year its hair was light brown, and its eyes were blue, but they had still, in a remarkable degree, though less so than before, that restlessness peculiar to albinos. This was the only case of the kind I ever heard of, except that communicated by Michaëlis in Blumenbach's 'Medicinische Bibliothek,' vol. iii. page 679; which, however, rests only on the uncertain authority of some peasants. Singularly enough," says Dr. Graves, "I had soon the good fortune to meet with a similar case myself. In my younger days there were two children, a brother and sister, living near me, who presented such striking symptoms of leucosis in their eyes, hair, and skin, that they were recognised as albinos even by non-medical persons. My attention was lately drawn to them by an advertisement in the papers in which their name occurred, and I learned that the brother had become a tobacconist; but, to my great astonishment, on going to see him, I found that his eyes had changed from violet-red to grey, and his hair from white to light brown, and that the susceptibility of the eyes to the light had greatly diminished."

The texture of the body in which all these varieties have their seat is the extracorial or exodermal structure, constituting, if I may so speak, the outer coating of the body, external to the true skin, which corresponds to the cuticular and corneous excrescences of animals,—a structure which includes horns, hoof, hair, feathers, and all similar appendages in different orders of animals. This structure displays infinite diversities in colour, constitution, and organisation, and is the most variable tissue on the whole body. Many different opinions have, however, been lately maintained, and much research has been made, as to the nature and texture of the parts on which the variety of colour depends; and it will be absolutely requisite, in order to obtain a tolerably correct knowledge of this subject, to trace briefly the history of these investigations.

The ancient anatomists were acquainted with two parts only of the common integument; they had no notion of anything interposed between the true skin, which is termed cutis, dermis, and sometimes corium, and the scarf-skin, or outer skin, called cuticle or epidermis. We may observe that these are the two principal constituent parts of the integument of the whole body, both in men and in all mammiferous animals. The epidermis is the outer covering of all the properly external surface of the body. When continued over the internal surfaces it is termed *epithelium;* while, by some, the name of epithelium is extended both to the cuticle and proper epithelium. The celebrated anatomist Malpighi was the first who discovered a third layer, or membrane, interposed between the cutis and the cuticle. He perceived that the seat of colour in the skin of the Negro is neither the epidermis nor the cutis, both of these parts of the skin being of the same colour as in the European. Malpighi had previously discovered a mucous membrane of a reticular texture, or consisting of a sort of net-work, situated beneath the epidermis in the tongue of an ox; and he conjectured the mucous membrane, which he afterwards found on the cutis in the Negro, to be a structure of a similar kind. From this supposition originated the term, long so popular, of *rete mucosum*. Albinus afterwards corrected the observations of Malpighi, and exhibited the coloured substance between the two skins as a continuous membrane. In his time the integument of the Negro was supposed to be, and was described in anatomical plates as consisting of, three distinct parts:—the cutis, white; the epidermis, of an ash colour; and the mucous layer, black. Long after this period, Cruikshank, in a series of observations on the skin of a Negro affected by the small-pox, discovered not less than four layers interposed between the cuticle and the true skin:—two situated beneath the colouring matter, the coloured layer itself, and a fourth exterior to the coloured layer. The investigation was pursued by M. Gaultier, principally by watching the effects produced by the application of blisters to the skin of a Negro. This writer also thought that he could discover four layers:—one consisting of vascular papillæ (*bourgeons vasculaires sanguins*), which has been termed the "corpus papillare," or papillary body; a second, which he names a deep-seated albugineous membrane; a brown substance, or layer of colouring matter; and a superficial albugineous membrane.

M. Flourens, who subsequently wrote an elaborate memoir on this subject, has attempted still greater precision. In his dissections he was enabled to display before the Academy of Sciences four distinct layers interposed between the outer cuticle and the cutis, without enumerating the papillary or vascular structure before alluded to. The account which this celebrated anatomist has given of his discoveries is extremely curious, and they have led him to important results. The four layers are, first, one lying immediately on the cutis, which is of cellular structure, and forms a reticular tissue; secondly, a continuous membrane which has the aspect of mucous membranes in general; on its external surface is spread the black pigment, which may be termed a layer, though it is not firm and coherent enough to be considered as a membrane,—this is reckoned as the third layer; exterior to the pigment is the fourth layer or membrane,—this is the interior portion of the epidermis, which is divisible into two distinct laminæ. Of these four layers the second, which is the mucous membrane, underlying the pigment, deserves the most particular attention, more especially as it constitutes, according to M. Flourens, a distinct organised body, which exists only in men of dark colour, and is entirely wanting in the white races, or at least, as M. Flourens says, "could not be detected in them by the *ordinary method of maceration*."* On the external surface of the mucous membrane is spread, as we have seen, the pigment, which, when denuded by maceration, is of much darker hue than when seen through the semi-transparent double cuticle. The internal surface of the same membrane is roughened by prolongations which pass through the interstices of the intervening cellular tissue, and fix themselves in the cutis. These prolongations form the sheaths of the hairs, reaching under their roots, and appearing to constitute the internal coat of the bulb of each hair: they only exist where the hairs arise. The mucous or pigmental membrane itself

* "Recherches Anatomiques sur le Corps Muqueux, ou Appareil Pigmental de la Peau, dans l'Indien Charrua, le Nègre et le Mulâtre." Par M. Flourens. Ann. des Sc. Naturelles. Paris. Tom. vii. p. 156.

is throughout of nearly equal consistence, and sufficiently thick to be divisible into two laminæ; it is on its outer surface that the colouring matter is spread. This matter, as we have observed, is not a distinct membrane; it is merely a layer or deposit, "*un enduit,*" a sort of plastering. It is immediately covered by a true continuous membrane, which is the inner lamina of the epidermis.

M. Flourens displayed by maceration all these layers in the skin of a Negro, in that of a Mulatto, and likewise in the integument of two Charruan Indians, belonging to a very dark race of native South Americans.* On subjecting the skin of a white man to the same process of maceration, he failed entirely to discover either the mucous membrane or the pigment which is deposited upon it. He was unable to find any membrane in the integument of a white man interposed between the white cutis and the outer lamina of the epidermis, except that internal coat of the epidermis itself already mentioned; and this, in his opinion, is the seat of the discoloration which is produced in the complexion of a white man by exposure to the heat of the sun.

M. Flourens is not the first anatomist who has attempted without success to discover the so-termed rete mucosum in the skin of white men. Dr. Gordon averred long ago, that after trying all the usual means, he was never able to succeed in tracing the existence of such a membrane in white persons. M. Flourens† has further maintained, that the seat of discoloration induced in the

---

* Two individuals of a now exterminated race, who were brought to France from their country near the Uruguay. I saw these persons living some years ago in Paris. Their complexion was as dark as that of many Negroes, and no epithet could be less appropriate to it than that of red, which is often applied to all the American tribes. A similar account is given of the colour of this tribe by Don Felix de Azara. They were a particularly ferocious, gloomy, and taciturn people, and, unlike their neighbours, the Guarani, were insusceptible of civilisation.

† "Recherches Anatomiques sur les Structures Comparées de la Membrane Cutanée et de la Membrane Muqueuse." Par M. Flourens. Ann. des Sc. Nat. tom. ix. p. 239.

white skin by exposure to heat is in the inner lamina of the cuticle; and, in a subsequent memoir, he pointed out the same membrane as the seat of the brown colour which is observed to surround the areola mammarum.

It was remarked long ago by Soemmering, that the cuticle is in the Negro of a browner and darker hue than in the European; but this does not coincide with the observation of M. Flourens, who considers the discolorations which take place from various causes in the skin of white men as totally different in kind, and having the seat in a different structure, from the cause of blackness in the Negro. The former change, according to him, depends merely on a hue imparted by temporary causes to the cuticle or scarf-skin, while the colour of a black man arises from a particular membrane entirely wanting in the white races. M. Flourens thus draws very strongly the line of discrimination between these two divisions of mankind; he regards the diversity in question as constituting a really specific distinction, or as marking out the Negro and European as separate species of beings. In fact, the endowment of an entirely peculiar organ to one race, of which no traces are to be found in the proximate tribe, is a much greater difference than is often to be found on comparing species which stand next to each other in the zoological series. Many facts were long ago on record in the works of medical authors, and others were known of almost daily occurrence, which are scarcely intelligible on the supposition maintained by M. Flourens. For example, a variety of disordered states of the constitution are known to occasion a very deep tinge of the skin in Europeans. Many females are well known to have a dark tinge extended over a considerable space round the mamma during the period of pregnancy, which disappears afterwards in a great measure. The change of colour which happens at such periods varies in its degree of intensity, as well as in the space occupied by it; and in some individuals it has been known to cover the abdomen, and even to affect the whole body. These facts are quite sufficient to prove that, independently of the influence of solar heat, a physical change may take place connected with the state of the constitution which imparts a black hue to the skin similar to that which is natural to the African race.*

This colouring matter is also liable to disappear by absorption in skins to which it is natural. Instances are not unfrequently observed in different countries, in which Negroes gradually lose their black colour and become as white as Europeans.†

It appears from these facts, that a substance giving occasion to various degrees of blackness is sometimes produced in the skins of persons born white, and that, on the other hand, the colouring substance in black persons is capable of disappearing; such phenomena appear insusceptible of explanation on the view of the structure of the skin adopted by M. Flourens, and they are calculated to direct our thoughts to the alternative which seems to be admitted by him, viz. that maceration and the ordinary purposes of examination by the natural eye afford insufficient means of ascertaining the intimate structure of the skin.

It is evident that microscopical investigation was alone

* Bomare, in an article cited by Blumenbach, mentions a French peasant whose abdomen became entirely black during each pregnancy; and Camper has given a particular account of a female of rank who had naturally a white skin and beautiful complexion, but whenever she became pregnant began immediately to grow brown. "Vers la fin de sa grossesse," he says, "elle dévenait une véritable négresse;" after delivery the dark colour gradually disappeared. Dr. Strach, in a work on intermittent fevers which is cited by Soemmering, mentions the case of a man who after a fever became as black as a Negro. Blumenbach says that he possesses a part of the skin taken from the abdomen of a beggar which is as black as the skin of an African. Haller, Ludwig, and Albinus have recorded similar instances.

† An example of this kind is recorded in the fifty-seventh volume of the "Philosophical Transactions." Klinkosch mentioned the case of a Negro who lost his blackness and became yellow; and Caldani declares that a Negro, who was a shoemaker at Venice, was black when brought during infancy to that city, but became gradually lighter, and had the hue of a person labouring under slight jaundice.

adequate to the solution of all these doubts, and to a display of the intimate structure of the tegumentary organs. This task has been commenced and performed in a satisfactory manner by the combined researches of several German anatomists, among whom Henle, Purkinje, and Schwann, are the most distinguished.* By these writers it has been proved that the outer integument does not consist at all of continuous membranes, but is of a cellular structure, and is composed of several layers of cells, and that its different parts are not distinguished from each other by such definite lines of separation as they have been supposed to be. The whole outer skin consists in reality of a complicated structure of cells, termed by anatomists cytoblasts, coating all the surfaces of the body. This exterior covering not only pervades the outer surface of the body, but is continued over all the mucous membranes into the excretory ducts, over the smooth and polished lining of the serous membranes, the inner surface of the heart, and that of the blood-vessels in their minutest ramifications. The cells, or cytoblasts, contain a solid nucleus of a round or oval form, and marked by one or two pointed granules. These nuclei are of uniform structure, but the transparent cells which envelope them vary in structure; and this variety constitutes the different appearances of the epithelia, or outer membranous linings of all the surfaces, of which there are, according to Henle, three kinds. The first kind, having the cells in close contact with the nuclei by which they are filled, are placed in a continuous structure; and from the resemblance of this to the stones placed in the pavement of a street, it is termed by Henle *Pflaster-epithelium*, or pavement-epithelium. This kind constitutes the covering of the cutis, and most of the serous membranes, including

* Henle, "Symbolæ ad Anatomiam Villorum Intestinalium, imprimis eorum Epithelii et Vasorum Lacteorum." Berol. 1837, 4to. Also "Ueber die Ausbreitung des Epithelium im menschlichen Körper," vom Prosector Dr. Henle, in Berlin. Müller's "Archiv," 1838, p. 103. Purkinje, Müller's "Archiv," 1836, p. 290. "Mikroskopische Untersuchungen," vom Dr. Th. Schwann. Berol. 1839.

the lucid cornea expanded over the ball of the eye. Cells of a conical form in different arrangements compose the two other kinds of epithelium, namely the cylindriform epithelium and the ciliated epithelium, which cover different internal surfaces of the body. In the pavement epithelium, or that coating which lies external to the cutis, the cells are seen to lie stratified one upon another, arising from the cutis, and undergoing changes of form according as they are compressed outwardly. More externally, the nuclei and the cells are progressively flattened, so that they appear at last as mere scales. The form of the cells, from being round, becomes polygonal from pressure, and finally, at the outer part, lamellar. In the outermost layers of the epidermis, the nuclei are scarcely at all discernible, and the laminæ, or scales, are so blended, that the aid of magnifying powers, all the previous steps of this gradual modification being known, is alone capable of displaying the real structure of the parts.

It thus appears that the idea of a given number of distinctly organised membranes, continuous and independent of the contiguous structures, must be abandoned: it was formed from an erroneous and defective view of the nature of the integumentary apparatus.*

Henle further examined the pigment membranes, as he terms them, namely, those apparently membranous parts which give a colouring to various surfaces; and he found that these are likewise of a cellular, and not properly of a membranous structure. The pigmental structure of the

* This view of the constitution of the skin is confirmed by the experiments of the most recent observers; in reference to this, I have been favoured with the following note by my friend, Dr. R. G. Latham:—"According to the evidence of the latest microscopists, the *rete mucosum* is no separate layer at all, but a part of the *epidermis*—of the epidermis as opposed to the corium or true skin. It exists equally in the white and dark families: in the latter, however, there is a deposit of colouring matter in its cells. It belongs to the epidermis, of which it is the innermost layer, lying immediately upon the *corium*. See 'Manual of Human Histology,' by A. Kölliker (Busk's and Huxley's Translation), vol. i. p. 132."—ED.

choroid coat of the eye is composed of polygonal cells, the centres of which are overlaid by colourless nuclei, and the surrounding space within the cells is partly filled by granules of pigment. The same anatomist made observations likewise on the skin of the Negro; and here he discovered, besides the order of cells above mentioned, others also, which contain the black pigment that tinges the African skin. He found these last aggregated especially on those parts of the rete which project and correspond with the furrows on the surface of the cutis. They resemble in form the cells in the pigment of the eye; that is, they are sometimes completely hexangular, but more commonly only approximating to this shape, being polyhedral, or irregularly spherical. Their length, according to Henle's measurement, reached to 0·0039—0·0062 line; their breadth to about 0·005 line.

Dr. Gustav Simon, of Berlin, at a somewhat later period, took pains to determine whether the various discolorations, or diversities of hue, which make their appearance occasionally on the skins of Europeans, including healthy and natural varieties of tint and those which occur in morbid states of the system, depend upon the presence of similar cells filled with pigment, or originate in some other manner.† To the natural or normal varieties of this description in the European skin belongs more especially that kind of discoloration which appears in the areola mammarum.* Dr. Simon says that he has frequently examined the areola in dead bodies, in examples in which the skin was there much discoloured, and decidedly of a brown hue. He found, on examining thin lamellæ, separated by means of perpendicular incisions, that the discoloration depended on the presence of cells filled with pigment. They lay in the rete Malpighii, and were seen in great numbers in the space between the papillæ of touch—*den*

---

* Müller's "Archiv," 1840. Heft 2, 180.

† P. 181. "Ueber die Structur der Warzen und über Pigmentbildung in der Haut," von D. Gustav Simon. Müller's "Archiv," 1840, 189.

*Gefühlswärzchen.* When he insulated the cells by tearing away a similar piece of the skin, only scraping off the epidermis, he immediately perceived clearly the pigment contained in them like small nuclei. Sometimes he perceived, also, cells which contained nuclei of pigment only in their periphery; while in their middle part none were to be seen. The shape and size of the particular cells agreed with Henle's description of the pigment-cells in the Negro. The proper epidermis appeared, both when the light fell upon it and when it shone through the transparent texture, *uncoloured;* and even when he soaked in water a piece of the skin from the areola so long that the epidermis began to get loosened, and examined one lamella of the outer skin after another, he could find no pigment-cells in the more superficial layers.

Besides the colouring of the areola mammarum, brown discolorations are often found, as it is well known, in other parts of the skin in Europeans. On examining the structure of the skin in these parts, Dr. Simon made observations precisely similar to those above related.

To the abnormal discolorations of the skin, which are produced by pigment-cells, belong especially many kinds of nævi materni, or congenital spots, as well as summer freckles. Of nævi materni, it is well known that there are two kinds, namely, vascular nævi and pigmental nævi. The latter form either spots greater in extent, and of various colour (these are the proper mother-spots so called), or present small brown, and sometimes even quite black spots which either do not at all raise themselves above the superficies of the skin, or, when otherwise, are seldom elevated more than the thickness of a pin, and are in general of an irregularly round form. These small round spots are called moles: they must not be confounded with the spots of ephelides, or pityriasis versicolor, which bear the same name.

Of the greater nævi materni Dr. Simon has examined two specimens, one of a dark brown, the other of a greyish black colour. The colouring material was also, in this case, contained in pigment-cells, which were situated in the rete Malpighii. They lay more thickly spread than those seen in the areola, as above described, but otherwise corresponded entirely with the cells found in that situation.

The moles of which Dr. Simon has been enabled to examine many, since they are often seen on bodies, are precisely similar in structure, when they are not elevated above the surface of the skin, to the two greater nævi above mentioned. In those which rise above the skin, the corium is found to form small projections, consisting of a cellular web imperfectly organised. On the surface of these the pigment-cells are dispersed, and over them the external skin. The latter appeared uncoloured, and in its superficial layers no pigment-cells could be perceived.

Even in summer freckles (*lentigo*) the colouring substance lies in the rete Malpighii, which, in places where these spots are found, appears of a clear brown when penetrated by light. When strong magnifying powers are applied, it is easy to be fully convinced of the presence of pigment-cells.

All these abnormal discolorations of the skin are therefore related, on the one hand, as Dr. Simon concludes, to the normal or natural colorations found in the Negro, and in many parts of the surface of the body of Europeans, and, on the other hand, make a transition to the character of the disease termed melanosis, in which, as J. Müller has proved, the production of pigment-cells keeps pace with a change from the normal or healthy state of organisation in the affected parts.

We may venture to conclude, from the results of these investigations, that there is no organic difference between the skin of the European and that of other races of men that gives reason to imagine a diversity of species in mankind; but, on the contrary, that transitions take place to a certain extent, independently of the agency of climate and the principal causes of variations, from the conditions of structure belonging to one race to those which characterise the other. We shall hereafter find reason to believe that this transition, under different circumstances, is very much more complete.

It will be worth while, before we take leave of this subject, to observe that the epidermic or horny tissue, corresponding in many tribes of animals to the extra-cutaneous texture which is the seat of variations in colour and in the hair of human beings, is precisely that part of the organic system which undergoes the most striking and even surprising alterations. It is this tissue which displays the variety of horns in tribes possessed of such appendages, some races of the same species having a great profusion of frontal antlers, while others are entirely destitute of them; and these variations, as we have seen upon evidence, are known actually to arise within the limits of one stock. The hoofs of many animals undergo similar changes: they are parts of the same structure. Perhaps of all instances of such deviation, that of the solid-hoofed swine is most remarkable, as there appears in this case an imitation of the really specific structure of other tribes of animals. Such a breed is well known; and nobody ever suspected it to constitute a distinct species.

It was long ago observed by Buffon that the skin itself displays great variations in many tribes of animals, and that this is the most palpable variation produced by the state of domesticity. The skin, for example, becomes softer in the domesticated race of asses; the wild asses of Persia have a rough tuberculated skin, which disappears on domestication. It is of the tuberculated skin of the wild ass that the Levantines make the grained leather termed *chagrin*.

The history of the "Porcupine Family" affords a curious illustration of the anomalies which display themselves in the textures external to the true skin, and of the wide range of variation to which these external coatings of the human body, as well as the corresponding parts in the inferior animals, are subject.

A boy, aged fourteen years, named Machin, born in Suffolk, was exhibited to the Royal Society in 1731. His body was covered by a remarkable kind of integument, which was thus described by Mr. Machin :—

"His skin, if it might so be called, seemed rather like a dusky-coloured thick case, exactly fitting every part of his body, made of a rugged bark or hide, with bristles in some places; which case covering the whole except the face, the palms of the hands, and soles of the feet, caused an appearance as if those alone were naked and the rest clothed. It did not bleed when cut or scarified, being callous or insensible. It was said he sheds it once every year, about autumn; at which time it usually grows to the thickness of three-quarters of an inch, and then is thrust off by a new skin which is coming up underneath." It was not easy to think of any sort of skin or natural integument that exactly resembled it. Some compared it to the bark of a tree; others thought it looked like seal-skin; others like the skin of an elephant, or the skin about the legs of a rhinoceros; and some took it to be like a great wart, or number of warts uniting, and overspreading the whole body. The bristly parts, which were chiefly about the belly and flanks, looked and rustled like the bristles or quills of a hedgehog shorn off within an inch of the skin. The second account of this person was communicated to the Royal Society by Mr. Baker. He was at that time forty years of age, and had been shown in London by the name of "the porcupine-man." He is described as a "good-looking, well-shaped man, of a florid countenance, who, when his body and hands are covered, seems nothing different from other people. But, except his head and face, the palms of his hands, and soles of his feet, his skin is all over covered in the same manner as in the year 1731; of which, therefore," continues Mr. Baker, "I shall trouble you with no further description than what you find in Mr. Machin's account above mentioned, only begging leave to observe that this covering seemed to me most nearly to resemble an innumerable company of warts, of a dark brown colour and a cylindrical figure, rising to a like height, and growing as close as possible to one another, but so stiff and elastic, that when the hand is drawn over them they make a rustling noise." "When I saw this man in the month of September, they were shedding off in several places, and young ones of a paler brown observed succeeding in their room, which he told me happens annually in some of the autumn or winter months; and then he is commonly let blood, to prevent some little sickness which he is subject to whilst they are falling off. He has had the small-pox, and has been twice salivated, in hopes of getting rid of this disagreeable covering; during which disorders the warting came off, and his skin appeared white and smooth like that of other people, but, on his recovering, soon became as it was before. His health at other times has been very good during his whole life.

"But the most extraordinary circumstance of this man's story is, that he has had six children, all with the same rugged covering as himself; the first appearance whereof in them, as well as in him, came on in about nine weeks after the birth. Only one of them is living, a very pretty boy, eight years of age, whom I saw and examined with his father, and who is exactly in the same condition. It appears, therefore, past all doubt," says Mr. Baker, "that a race of people may be propagated by this man, having such rugged coats and coverings as himself; and if this should ever happen, and the accidental original be forgotten, it is not improbable they might be deemed a different species of mankind."

That the different complexions of mankind are not permanent characters may be sufficiently proved by numerous facts collected from the physical history of particular races of men. It is hardly necessary, in this instance, to appeal to the infinite number of phenomena which are to be found, precisely analogous in all the circumstances of their origin and subsequent propagation and permanence in entire breeds, in the various tribes of animals, there being scarcely any tribe of warm-blooded creatures which is not subject to become thus diversified. The reader will find, in the following outline of the history of particular tribes of the human family, instances of this variation of colour,—of a change from white to black, and from black to white, or of both complexions actually subsisting in the undoubted progeny of the same stock; and these instances so multiplied and so well authenticated, as to leave no doubt as to the conclusion which we are obliged to draw in this part at least of the investigation before us, as to the great question of the unity or diversity of the human species.

## CHAPTER XI.

### ON THE STRUCTURE OF THE HUMAN HAIR.

The structure of the hair has not been so fully elucidated as that of the skin, though much labour has been bestowed upon the subject by various microscopical observers since the elaborate works of Heusinger and Eble were published.* These writers supposed the human hair, like quills of hedgehogs and porcupines and the bristles of hogs, to consist of two distinct parts, a cortical outer coat, and an internal spongy structure. This has been denied by later observers: Weber declares the human hair to consist of a homogeneous substance, in which no distinction of cortex and medulla can be perceived.

The hair issues from follicles, by a club-shaped root or bulb, compared to a bud (*keim*), which is thicker than the filament of the hair. Müller supposes that the substance of the hair is formed by the secretion of a horny matter from the surface of a conical vascular pulp, contained

* "System der Histologie," v. Heusinger. Eisenach, 2, 1823. "Die Lehre v. der Haaren," v. Dr. Burkard Eble. 2 Bänden in 8vo. Wien, 1831. See Quekett, "Lectures on Histology," 2 vols. 8vo. London, 1851-54.

within the bulb, and which, in fact, is a vascular prolongation of the bottom of the follicle.* The hair, as he says, grows by the addition of new matter at its root; consequently the extremity is the part first formed.

Since the investigations of Henle and Schwann have demonstrated the cellular organisation of the skin, some researches have been made as to the hair, in the expectation of detecting a similar structure in its component parts. A paper of Gürlt's in Müller's "Archiv"† had this purport; and the subject has been more recently pursued in a memoir by Dr. Bidder of Dorpat, published in the same journal.‡

At the origin of each hair, two parts, according to Dr. Bidder, are distinguished: the hair-sheath (*haarbalg*), and the germ, or hair-bud—*bourgeon*—*haarkeim*. The bud reaches below the commencement of the sheath, and at its extremity, where it is joined to the surrounding soft parts, is, even as perceived by the naked eye, of an intensely dark colour. The extremity of the hair appears under the microscope a dark-coloured mass, consisting of small grains, which can be separated by the aid of acetic acid and careful division, and be shewn to consist of minute but distinct cells, or cytoblasts, each containing its nucleus. The cells are united by a tenacious inorganic substance, to which the name of cytoblastema is given. The sheath which envelopes the substance of the hair is lined within by a peculiar epithelium, consisting of colourless and transparent cells: into this structure the cellular one of the bud passes uninterruptedly, but the latter is distinguished by its dark colour. The cells of the bud take the form of extremely fine threads, and appear like fibres lying parallel to each other, and united by a transparent cytoblastema. The entire hair thus may be said to consist

* Müller's "Physiology," translated by Baly, vol. i. p. 398.

† Gürlt, in Müller's "Archiv," 1836.

‡ "Bemerkungen über Entstehung, Bau und Leben der menschlichen Haare," von Dr. Bidder, in Dorpat. Müller's "Archiv," 1840.

of a bundle of longitudinal fibres. The origins of these are threadlike cells, which are continued from the bottom of the hair-bud to the end of the hair itself, but which, in this course, undergo essential changes in size and form. The developed hair is uniform through its whole mass; no pith and external coating can be distinguished; but, by maceration in concentrated acids, it becomes so soft that it can be separated into longitudinal threads; but these threads are found under the microscope to be bundles of numerous fine fibres. The latter having an extremely small diameter, are the elementary constituents of the hair: they have the appearance of dark lines, becoming in some places broader than others, united by a yellowish cytoblastema. The thickness of a human hair from the head being estimated at one-tenth of a line, there must be 250 of such elementary fibres, without reckoning any space for the certainly existing cytoblastema, to fill up this extent.* But in this part of Dr. Bidder's investigation some difficulty arises as to the size and number of these fibres, compared with the dimensions of cells in the bud from which they originate; and this he confesses that he has not been able to clear up.

As to the seat of colouring matter in the hair, Dr. Bidder concludes that it is produced in the lowest parts of the bud by the dark contents of the cells. It is, therefore, precisely analogous to the colouring of the rete mucosum. But in the developed hair, it appears, as he says, that the chief seat of colour is in the yellowish or brownish-coloured mass, termed cytoblastema, which surrounds the fibres, and is exterior to them.

It is to be hoped that on this and other subjects connected with the structure of the hair, further elucidation is hereafter to be obtained from microscopical researches.

* Dr. Grant has remarked that this space, filled by the 250 fibrils, is the mere diameter of the hair, and that to fill its entire calibre there must be about fifty thousand.—See Dr. Grant's "Outlines of Comparative Anatomy," p. 647, 8vo. with 147 woodcuts. London, 1841.

The probable conclusion from all that is at present known seems to be, as it has been extremely well expressed by an ingenious and able writer, who has illustrated many parts of minute anatomy by comparing analogous structures in different orders of beings, "That each of the cells contained in the hair-bud gives origin to a bundle of fibres, in the same manner as does that of the cortical substance of the feather, and that the fibres are really, in both instances, elongated secondary cells."*

### *Of the National Differences of the Human Hair.*

The varieties as to colour and structure in the hair of different nations is one of the most remarkable diversities of physical character that distinguishes them.

The colour of the hair of animals varies with climate. Eble observes, that in northern regions there are no black horses; and that hares, squirrels, weasels, and many other animals, are there white. This does not hold true with respect to all tribes of animals, as the sables brought from Siberia testify. In the human kind it is true in a general point of view, but with many exceptions. The facts connected with this part of the subject will hereafter come under our observation. With respect to the quantity of hair that grows on the human body there are well-known differences between races: the Mongoles, and other Northern Asiatics who are similar to them, are noted for the deficiency of their hair and for scanty beards; and the same character is ascribed to all the American nations, who, in some other respects, resemble the Northern Asiatics. Blumenbach and Eble conjectured that the habit of pulling out the hair, continued through many generations, may have produced at length this national variety. But it is too general to be ascribed to so accidental a cause.

We find some instances of races bearing an exuberant

---

* "Principles of General and Comparative Physiology," by Dr. W. B. Carpenter. 2d edit. London, 1841.

growth of hair. Among the *Ainos*, or in the Kurilian race, there are individuals who have hair growing down the back and covering nearly the whole body.

It is probable that none of these national diversities exceed that measure of variety which occurs in the same nation in different families.

The Northern Asiatics and Americans have generally straight lank hair, though with occasional exceptions. Europeans have it sometimes straight and flowing, at others considerably curled and crisped; I have seen some Europeans whose hair is nearly, if not quite, as crisp as that of a Negro. Even among Negroes themselves there is a very great variety; and if we take the entire mass of the black native races of Africa into comparison, we shall find tribes among them who, similar in complexion and in most other physical peculiarities, yet differ in regard to their hair, and present every possible gradation, from a completely crisp, or what is termed woolly hair, to merely curled, and even to flowing hair. A similar observation holds respecting the natives of the islands in the great Southern Ocean, where some individuals have crisp, and others merely curled hair; this variety occurs in the same race, and where there is no reason to suspect intermixture of breeds. These examples will come under our notice. It has been observed that the hair on the eyebrows and eyelids of Negroes is not woolly in appearance, though more curled than in Europeans. The nature of the crisp, and, so termed, woolly appearance of the hair in the Negro, must be made a subject of careful investigation, especially as this is one of the characters which give rise to a suspicion of distinctness of race and origin.

### *Of the Nature of the Hair of a Negro.*

The hair of the Negro has been thought the most decidedly different from that of other human races. It is commonly said that the substance which grows on the head of the African races, and of some other dark-coloured tribes, chiefly inhabiting tropical climates, is wool, and not hair. In order to determine this point, it would be requisite to form a clear idea of the difference between hair and wool.

Dr. Eble examined with a microscope the wool of the Merino and of the Chinese sheep, and found a striking difference between these substances. He says, that all wool displays filaments twisted and matted in all directions; and, moreover, the shaft of the filament of wool does not keep an uniform calibre, but appears thickened here and there, and often swelled out with an appearance of knots. He adds, "I could everywhere distinguish clearly the so-termed medulla or pith—the transparent canal; and could accurately discriminate between this and the cortical substance. Yet in the various proportions which these parts bear to each other is to be found the chief difference between the finest and the coarsest wool. The cortical part appeared in both almost equally thick and intransparent, at least near the edge; but the canal in common wool seemed divided into more numerous spaces, resembling irregular cells, while in the wool of the Merino sheep the cells appeared more regularly arranged. The whole canal of the sheath seems to be separated by regularly interposed fine transverse laminæ. The hair of the Tibetan goat, of which the Kashmire shawls are made, approaches in its texture very nearly to the Merino wool, only it is smaller in the diameter of the sheath, and the transverse laminæ appear not so regularly placed. The Chinese sheep has wool, and interspersed among it rough coarse hair."*

It has been conjectured by Mr. Monge and others, that the felting quality of wool is owing to the rough nature of the surface of its filaments, and that these filaments have a feathered or barbed edge; but the merit of discovering this property as a matter of fact is due to Mr. Youatt. According to this writer, the true cause of the felting property of wool, and, at the same time, the character which

* "Die Lehre von der Haaren," von Dr. Eble. Th. 1.

distinguishes it from hair, consists in the serrated nature of its external surface. When examined through a microscope of great power, the fibre of Merino wool assumes a riband-like form with serrated edges. When the fibre is viewed as an opaque object, the serrations are found to result from a structure resembling a series of inverted cones, encircling a central stem, the apex of one cone being received into the base of the superior one; each cup-like cone having indented edges directed from root to point, as seen in the annexed figures.

FIG. 23.

A fibre of long Merino wool, viewed as a transparent object.

FIG. 24.

The same viewed as an opaque object.

Hair, although sometimes covered with scales or rugosities, has no serrations or tooth-like projections. The hair of a tiger is covered with scales like those on the back of a sole; while in the wool of the same animal the serrations are distinct and numerous.

FIG. 25.

The wool of the Tiger, transparent.

FIG. 26.

The hair of the Tiger, transparent.

The wool of a rabbit is fine, with sharp angular serrations to the number of 2880 in an inch. The hair of a

FIG. 27.

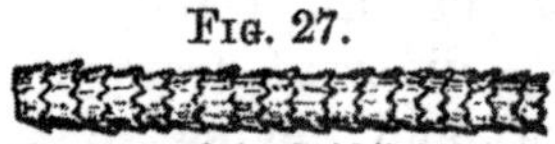

The wool of the Rabbit, opaque.

rabbit varies from the $\frac{1}{250}$ to the $\frac{}{300}$ part of an inch in diameter, and is covered with a scaly incrustation, which cannot be said to be serrated.

FIG. 28.

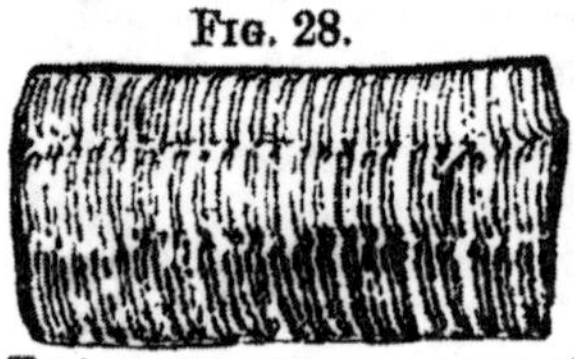

The hair of the Rabbit, transparent.

FIG. 29.

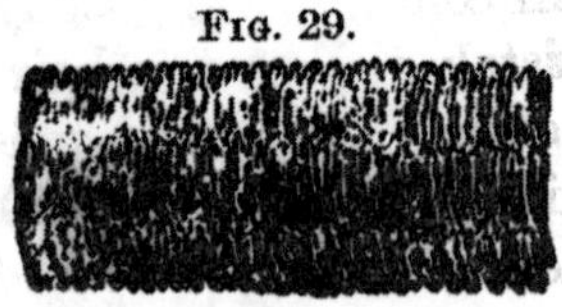

The hair of the Rabbit, opaque.

The wool of a bear, which is very fine, has serrations which, as Mr. Youatt says, resemble so many spines projecting at irregular distances, and at a very acute angle. In the Italian wolf-dog, which has a considerable portion of short wool beneath the hair, the serrations of the wool were found by Mr. Youatt to be superficial and irregularly placed, some of them resembling small spines, and others like rounded prominences. When viewed as an opaque object, the cup appeared to be composed of two or three leaves with rounded extremities.

FIG. 30.

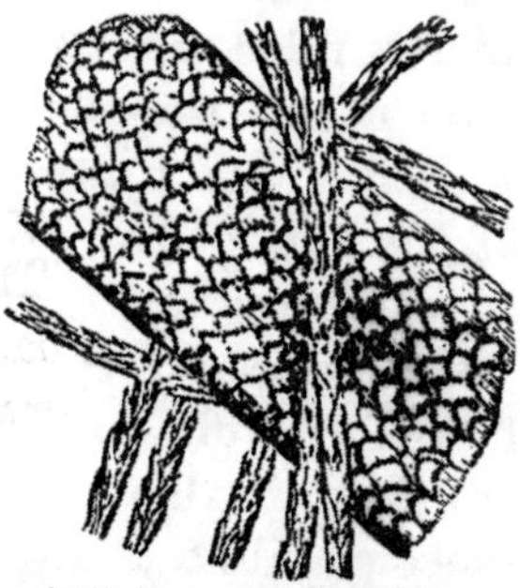

Seal's hair, *a*, and wool, *b*.

FIG. 31.

The wool of the Bear.

FIG. 32.

The wool of the Wolf-Dog, transparent.

FIG. 33.

The wool of the Wolf-Dog, opaque.

It appears certain, from the observations of Mr. Youatt, that Merino wool, and some other kinds of wool, consists of filaments having serrated edges. But there are many kinds of wool produced by other breeds of sheep, in which this property cannot be discovered. In these, however, the filament is very different from that of hair, being of unequal thickness, and having rough, uneven edges; whereas the filament of hair is a smooth and even-sided tube, and nearly of equal calibre.

A careful observation, with the aid of the microscope, will convince every body who makes it, if I am not much mistaken, that the hair of the African is not wool, but merely a curled and twisted hair. I have seen and examined the filaments of hair belonging to different races of men, and have compared them with the filaments of wool from the Southdown sheep, with the assistance of Mr.

Estlin, who is skilful and long practised in the use of the microscope, with the aid of glasses magnifying about 400 times. Hairs of a Negro, of a Mulatto, of Europeans, and of some Abyssinians sent to me by M. d'Abbadie, the celebrated traveller, were, together with the wool of a Southdown sheep, viewed both as transparent and opaque bodies. The filament of wool had a very rough and irregular surface, though no serrations, distinctly so termed, were perceptible. The filament of Negro's hair, which was extremely unlike that of wool, and of all the other varieties mentioned, had the appearance of a cylinder with smooth surface; they all appeared more or less filled with a dark colouring matter, which, however, did not entirely destroy their transparency. The colouring matter was apparently much more abundant in the hair of the Negro than in the others. The Abyssinian hair was also very dark, but so far diaphanous that a riband-like band appeared running down through the middle of a cylindriform tube; and the Mulatto hair resembled the Abyssinian in this respect. The filament of European hair seemed almost entirely transparent; it had the appearance of an empty tube, coated internally with something of a dingy or dusky colour, which only prevented it from being quite pellucid. European hair of a light colour had the same appearance, but was still less darkened.

From these observations, I am convinced that the Negro has hair properly so termed, and not wool. One difference between the hair of a Negro and that of a European consists in the more curled and frizzled condition of the former. This, however, is only a difference in the degree of crispation, some European hair being likewise very crisp. Another difference is the greater quantity of colouring matter or pigment in the hair of the Negro. It is very probable that this quality is connected with the former, and is its cause, though we cannot determine in what manner one depends upon another; but as these properties vary simultaneously, and are in proportion one to another, we may infer that they do not depend upon independent causes.

It may be worth while to remark, before we take leave of this subject, that if this cuticular excrescence of the Negro were really not hair, but a fine wool,—if it were precisely analogous to the finest wool,—still this would by no means prove the Negro to be of a peculiar and separate stock, since we know that some tribes of animals bear wool, while others of the same species are covered with hair. It is true that in some instances this peculiarity depends immediately on climate, and is subject to vary when the climate is changed; but, in others, it is deeply fixed in the breed, and almost amounts to a permanent variety.

## CHAPTER XII.

### OF THE VARIETIES OF FORM OR CONFIGURATION—AND OF THE SUBDIVISION OF RACES INTO PARTICULAR GROUPS.

By most writers on the diversities of mankind, the varieties which are observed in the form and structure of the internal parts, and particularly in the bony fabric, including the skull, are looked upon as furnishing the most important characters of different races, and as constituting those marks of separation which have the best claim to be reckoned specific distinctions. Differences in the form of the body, in the proportion of parts, in the size of the head, and the development of the brain, have been thought to be more essential and important characters than the external phenomena relating to the complexion, or the texture of the skin and hair: it has been supposed that they are liable to fewer irregularities or anomalous changes.

Varieties of form and structure depend chiefly on differences in the bony fabric; and, among the differences of the bony system, none are more striking than those which have been observed in the shape of the skull. Accordingly, since the time of Camper and Blumenbach, various attempts have been made by anatomists to divide mankind into groups, by taking the shape of the skull as the chief ground of distinction. Some authors have made but a few departments, others many; scarcely any two writers are fully agreed as to their manner of distribution. In one particular most have erred: it has been generally laid down as a fundamental principle that all those nations who are found to resemble each other in the shape of their heads must needs be more nearly related to each other than they are to tribes of men who differ from them in this particular; and they are accordingly set down as constituting so many different races or families. This would be apparently true if it were established that all the organic differences observed in mankind are absolutely permanent, and are, in fact, the distinguishing marks of separate species. But while it is still allowed that they may be merely varieties, which, for aught that has been proved to the contrary, may have been produced by external agencies on the different branches of one original stock, it must be considered probable that similar causes may have produced on many different tribes similar effects; and the inference is, that a mere resemblance in some particular anatomical characters affords no infallible proof of near relationship. When we discover in different parts of the world tribes of men who are similar in the shape of their heads, or in any other particular, we may not at once determine that they belong to one race, or are nearly connected in origin.* In distributing the varieties of form and of anatomical structure found in mankind into several departments, the object to be kept in view is to assist our estimate of the extent of variation in particular divisions of the human family, and the comparison of the most

* Thus, when Barrow concluded, chiefly from some resemblance in the shape of the head, that the Hottentots are the descendants of the Chinese, he evidently drew an inference which requires further proof.

different tribes. Hence it is of no essential importance how many or how few of such groups are enumerated.

If any method of subdividing the human family into groups is likely to be of any particular advantage in elucidating the natural history of the species, it must be one founded on some relation between the physical characteristics of different tribes and the leading circumstances of their external condition. We shall clearly perceive, in tracing the following outline of ethnography, that the varieties of colour refer themselves, in part, to climates, elevations of land, proximity to the sea-coast, or distance from it. It can hardly be doubted that these conditions have likewise an effect on the configuration of the human body. But there is, perhaps, some truth in the remark, though frequently made on little better foundation than conjecture, that the prevailing form or configuration of the body is more liable to be influenced by the habits of different races and their manner of living than by the simple agencies of climate. It would be an interesting discovery, could it be shown that there is any apparent connection between the display of particular forms, or the leading physical characters of human races, and their habits of existence. If I may venture to point out any such relation, it would be by remarking, in a very general manner, and without pretending to make the observation as one which holds without many exceptions, that there are in mankind three principal varieties in the form of the head and other physical characters, which are most prevalent respectively in the savage or hunting tribes, in the nomadic or wandering pastoral races, and in the civilised and intellectually cultivated divisions of the human family. Among the rudest tribes of men, hunters and savage inhabitants of forests, dependent for their supply of food on the accidental produce of the soil, or on the chase, among whom are the most degraded of the African nations and the Australian savages, a form of the head is prevalent which is most aptly distinguished by the term prognathous, indicating a prolongation or extension forward of the jaws; and with this characteristic other traits are connected, which will be described in the following pages. A second shape of the head, very different from the last mentioned, belongs principally to the nomadic races, who wander with their herds and flocks over vast plains, and to the tribes who creep along the shores of the Icy Sea, and live partly by fishing, and in part on the flesh of their reindeer. These nations have broad and lozenge-formed faces, and what I have termed pyramidal skulls. The Esquimaux, the Laplanders, Samoiedes, and Kamchatkans, belong to this department, as well as the Tartar nations, meaning the Mongolians, Tungusians, and nomadic races of Turks. In South Africa, the Hottentots, formerly a nomadic people, who wandered about with herds of cattle over the extensive plains of Kafirland, resembling in their manner of life the Tungusians and the Mongoles, have also broad-faced, pyramidal skulls, and in many particulars of their organisation resemble the Northern Asiatics. Other tribes in South Africa approximate to the same character, as do many of the native races of the New World.

The most civilised races, those who live by agriculture and the arts of cultivated life, all the most intellectually improved nations of Europe and Asia, have a shape of the head which differs from both the forms above mentioned. The characteristic form of the skull among these nations may be termed oval, or elliptical.

We shall find hereafter that there are numerous instances of transition from one of these shapes of the head to another, and that these alterations have taken place in nations who have changed their manner of life. I shall only mention one example at present: the nomadic tribes of Turks spread through Central Asia have the configuration of skull mentioned in the second place in a very marked degree. The long-since civilised Turks, descended from the early conquerors of Maweralnahar and Khorasan, and the Seljukians, who for eight centuries have inhabited the Ottoman and Persian empires, have become completely transformed into the likeness of Europeans. Some have attributed this change in the physical structure of the Turkish race to the introduction of Circassian slaves into the harems of the Turks, but this could only affect the opulent and powerful among the race: the great mass of the Turkish population have always intermarried among themselves, and the difference of religion and manners must have kept them separate from those Greeks whom they subdued in the new Ottoman countries; while in Persia, the Tajiks, or real Persians, belong to a different sect of Musulmans, and are still a distinct people from the Turks who govern them, and who inhabit much of the open country remote from towns.

I shall now give some further account of the varieties above noticed in the shape of the skull, and of the methods of investigation adopted by the most celebrated writers who have entered on this subject.

## CHAPTER XIII.

### OF THE PRINCIPAL FORMS OF THE SKULL—AND OF THE MODES OF MEASURING IT ADOPTED BY ANATOMISTS.

THE prognathous form of the skull is most strongly developed in some of the tribes of Western Africa. It prevails in a less degree in many of the native African tribes who are commonly termed Negroes, but is by no means universal among the nations so named, if the designation is used as comprehending all those people of the African Continent who combine frizzled and crisp hair with a sable skin. Prognathous skulls are found also in the Eastern Ocean: the Pelagian Negroes of the great Austral islands, as well as the Alfurian* or Australian races, have the general form of their skulls of a similar description, though in other respects different from the prognathous

* See note in page 21.—ED.

heads of the African nations. I shall describe these varieties in the sequel; I refer at present to Negroes of Guinea or Western Africa. The prognathous form is most strongly marked in the tribes inhabiting the country between the long chain of the Kong Mountains and the sea-coast stretching east and west from Cape Palmas to the bottom of the Bight of Benin.

It has often been said that the form of the head in the Negro makes some approach to that of the chimpantsi and other simiæ. This is true to a very slight degree; but the resemblance consists in the greater extension of the jaws; it has no relation to the cranium, properly so termed, or the bony apparatus which contains the brain. I shall take, however, some account of the skulls of the higher simiæ in comparison with the human skull, as illustrating the nature of this peculiarity, which the orang and chimpantsi have in a much greater degree.

There are three ways of viewing the form of the skull, which, taken jointly, enable us to form a correct idea of the whole of its characters. These methods have been pointed out by three different authors, who have successively directed their attention to the subject. The lateral view of the skull was first described by Professor Camper, who measured the profile by his celebrated facial lines: the vertical outline, or the configuration of the skull, and the measurement of its area, when we look down upon the vertex, was the aspect which seemed most important to Professor Blumenbach: lastly, Professor Owen first pointed out the importance of comparing the figures given by the basis of the skull, or the under surface of the cranium, the lower jaw being removed. We must take all these different views into our account, in order to obtain a complete idea of the character of the head for the comparison of human races.

### 1. *Lateral View of the Skull or Profile: Facial Lines of Camper.*

Camper was the first anatomist who attempted to distinguish and describe in an accurate manner the differences of form which have been discovered on comparing the skulls of different human races. This writer invented a

FIG. 34.

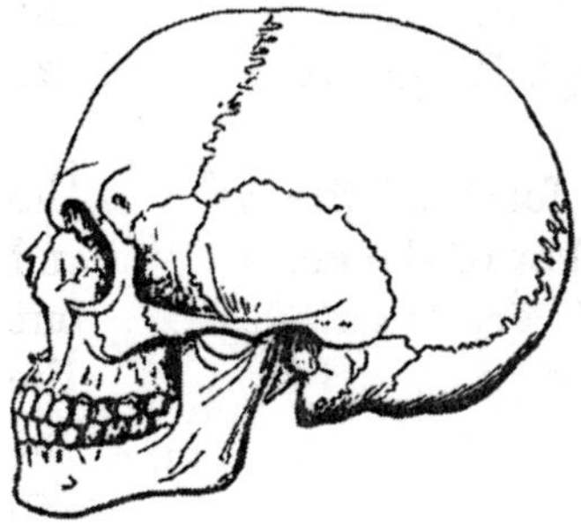

European.

FIG. 35.

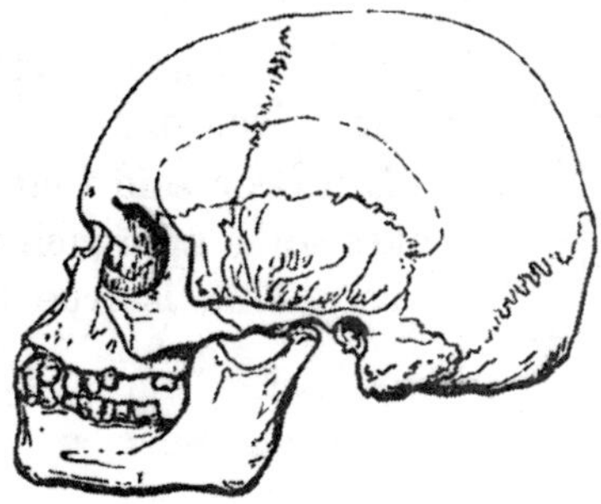

Negro.

FIG. 36.

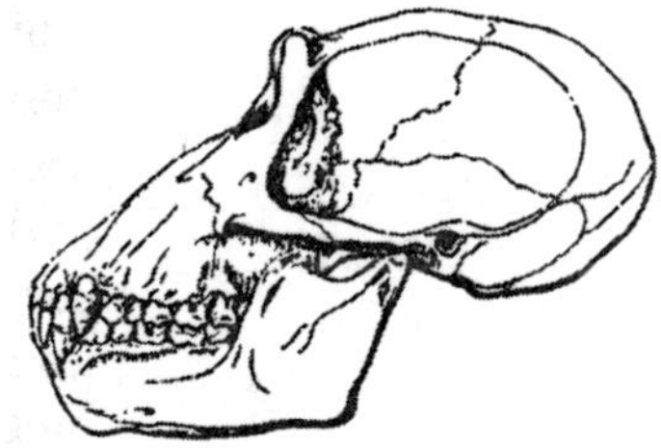

Chimpantsi.

FIG. 37.

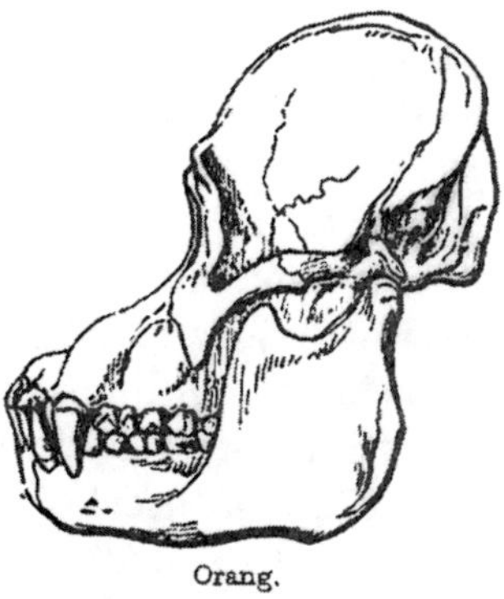

Orang.

technical method, by which he imagined that he could display, in a single measurement, the essential difference of skulls as to form and capacity, not only in reference to various nations of men, but likewise as to the inferior species of animals. His own account of this method is as follows:—

"The basis on which the distinction of nations is founded may be displayed by two straight lines; one of which is to be drawn through the meatus auditorius to the base of the nose, and the other touching the prominent centre of the forehead, and falling thence on the most advancing part of the upper jaw-bone, the head being viewed in profile. In the angle produced by these two lines may be said to consist, not only the distinctions between the skulls of the several species of animals, but also those which are found to exist between different nations; and it might be concluded that Nature has availed herself, at the same time, of this angle to mark out the diversities of the animal kingdom, and to establish a sort of scale from the inferior tribes up to the most beautiful forms which are found in the human species. Thus it will be found that the heads of birds display the smallest angle, and that it always becomes of greater extent in proportion as the animal approaches more nearly to the human figure. Thus there is one species of the ape tribe in which the head has a facial angle of forty-two degrees; in another animal of the same family, which is one of those simiæ most approximating in figure to mankind, the facial angle contains exactly fifty degrees. Next to this is the head of the African Negro, which, as well as that of the Kalmuk, forms an angle of seventy degrees; while the angle discovered in the heads of Europeans contains eighty degrees. On this difference of ten degrees in the facial angle the superior beauty of the European depends; while that high character of sublime beauty which is so striking in some works of ancient statuary, as in the head of Apollo, and in the Medusa of Sisocles, is given by an angle which amounts to one hundred degrees."

The theory founded by Camper on this measurement, of a gradation in different orders of beings, in which the Negro forms an intermediate step between the European and the orang, has been, as far as it regards the human skull, entirely overthrown by the curious and interesting discoveries of Professor Owen. It must be observed that Tyson, Camper, and earlier anatomists who have written on the structure of the simiæ, founded all their observations on orangs of immature age; hence their remarks on the facial angle, teeth, and the relative proportions of the cranium and the face, are erroneous when applied to the adult animal, and have led, as Mr. Owen has clearly proved, to an opinion that the transition from mankind to the simiæ is much more gradual than it really is. It is well known that in the immature and undeveloped state anatomical relations are, in many instances, nearer than they appear when the entire being is perfected, and prepared for all the functions for which Nature has destined it. Thus the human fœtus is well known to have a separate intermaxillary bone, in common with the simiæ and other inferior animals, while the absence of this separate strucure in man has ever been regarded as one of his distinctive anatomical characters. It is no matter of surprise, that when the skull of the young chimpantsi was examined at the period when the small deciduous teeth only are developed, the resemblance to the human cranium should have been found surprisingly close. The brain in the ape attains its full size at a very early period: it is not destined for further development, like the human brain; consequently, at the age when the jaws become enlarged, and lengthened with the increase of the maxillary apparatus, and the zygomatic arch is extended without any corresponding downward growth and development of the brain, or extension of its containing cavity, the proportions of the cranium to the jaws undergo a material change. In the earlier period, when its cranial portion preponderates over the facial and maxillary part, the head of the orang approximates to the human form; the facial angle is wide; the occipital foramen is more central; and the zygomatic arches, when the basis of the skull is examined, appear confined to the anterior half of the cranium. All these characters of resemblance are surprisingly changed when adult skulls are compared; it then appears, as Mr. Owen has shown, that strongly marked and most important characters distinguish the heads of quadrumanous animals from those of human beings. The cranium, properly so termed, is a small rounded case, and is altogether posterior to, and not above, the face.

The importance of this difference of age with respect to the facial angle is in the simiæ very great. In Camper's measurement this angle extends, in the heads of Europeans, to eighty degrees. In some human skulls it has been considerably less, and has even measured only seventy degrees, according to the same writer, in the heads of Negroes. In the orang, it has been estimated at sixty-four degrees, sixty-three degrees, or sixty degrees; but this has been in the measurement of the skulls of young apes. It is stated by Mr. Owen that the facial angle of the adult troglodyte is only thirty-five degrees, and that of the orang, or satyr, thirty degrees.

Such being the extent of difference manifest between the skulls of those simiæ which most approach the human form,—a difference so great that the utmost diversity between human races is quite inconsiderable when compared with it,—it becomes rather a matter of curiosity than of importance to the solution of any great question, to know whether the cranium of any human tribe actually makes a slight advance towards the type of the troglodyte or orang. Still, the facts which Soemmering and other anatomists have pointed out are not to be overlooked.

### *Of the Vertical Configuration of the Skull.*

The four figures placed on the opposite page will sufficiently display the vertical method—*norma verticalis*—adopted by Blumenbach as the principal measurement in the comparison of skulls. The four figures are those of a European skull; a Mongolian, or Tartar; a Negro; and fourthly, the artificially elongated skull of the ancient Peruvian, found in the burial-places at Titicaca by Mr. Pentland.

Blumenbach gave the following account of this way of describing heads, which, he says, is the result of his own observations in a long and constant study of his collection of the skulls of different nations. He remarks, that the comparison of the breadth of the head, particularly of the vertex, points out the principal and most strongly marked differences in the general configuration of the cranium.

FIG. 38.

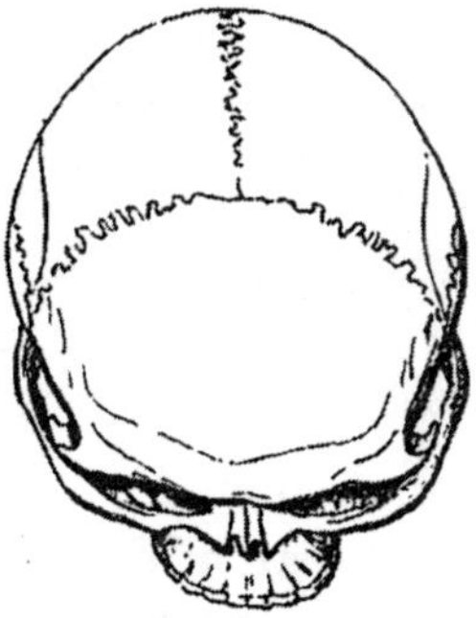

Caucasian.

FIG. 39.

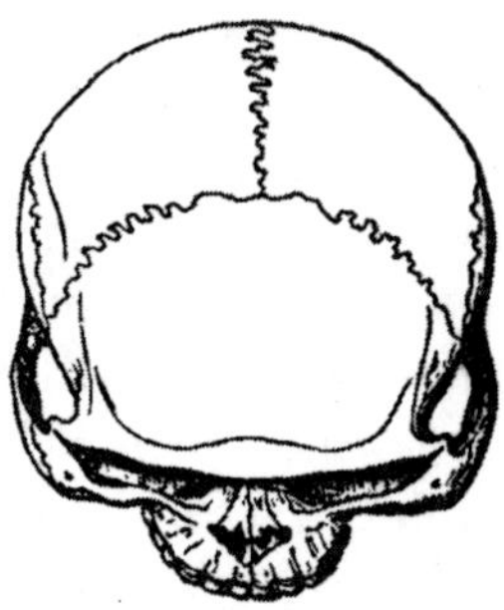

Mongole.

FIG. 40.

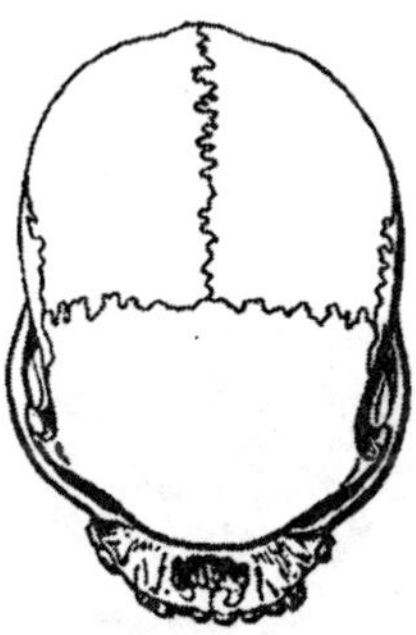

Negro.

FIG. 41.

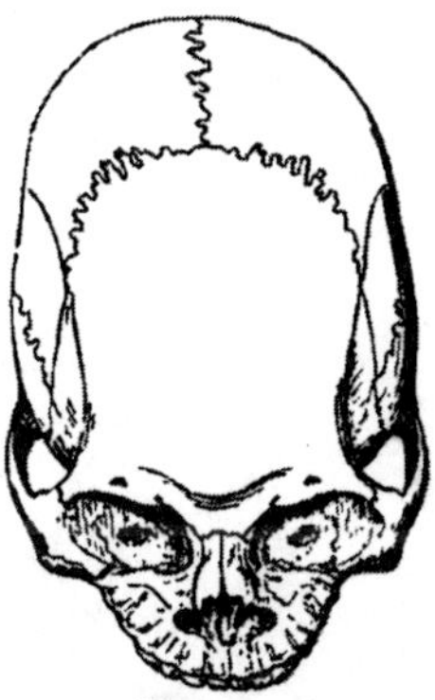

Titicacan.

He adds, that the whole cranium is susceptible of so many varieties in its form, the parts which contribute more or less to determine the national character displaying such different proportions and directions, that it is impossible to subject all these diversities to the measurement of any lines or angles. In comparing and arranging skulls according to the varieties in their shape, it is preferable to survey them in that method which presents at one view the greatest number of characteristic peculiarities. "The best way of obtaining this end is to place a series of skulls with the cheek-bones on the same horizontal line resting on the lower jaws; and then, viewing them from behind and fixing the eye on the vertex of each, to mark all the varieties in the shape of parts that contribute most to the national character, whether they consist in the direction of the maxillary and malar bones, in the breadth or narrowness of the oval figure presented by the vertex, or in the flattened or vaulted form of the frontal bone."

When all the different forms of the human cranium are compared with each other in the way thus pointed out, there are, as Blumenbach observes, three varieties in the vertical figure strongly distinguished from each other. The skulls of a Georgian, a Tungusian, and a Negro of Guinea, are given by him as specimens of the three varieties of form which he terms Caucasian, Mongolian, and Ethiopian.

*Measurements of the Basis of the Skull.*

No single view of the skull determines so much in regard to its general configuration as that of the basis.

FIG. 42.

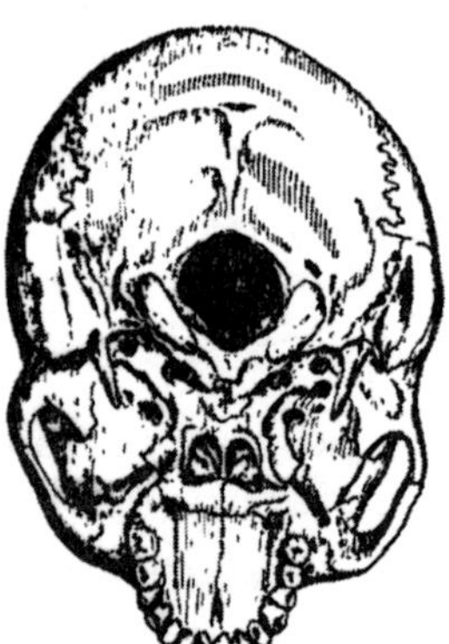

Base of Human Skull.

FIG. 43.

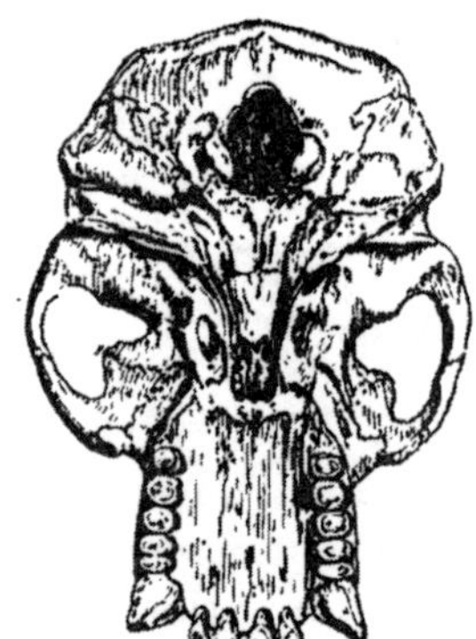

Base of Skull of Orang.

The importance of this manner of examining the bony structure of the head has been demonstrated in the fullest manner by Mr. Owen, in his excellent memoir on the structure of the orang and chimpantsi. The relative proportions and extent, and the peculiarities of formation of the different parts of the cranium, are more fully discovered by this mode of comparison, which has hitherto been much neglected, than by any other method.

It may be observed, in this view of the cranium, that the antero-posterior diameter of the basis of the skull is in the orang very much longer than in man. The most striking circumstance which displays this difference is the situation occupied by the zygomatic arch in the plane of the basis of the skull. In all races of men, and even in human idiots, the entire zygoma is included in the anterior half of the basis cranii; in the head of the adult troglodyte or chimpantsi, as well as in that of the satyr or orang, the zygoma is situated in the middle region of the skull, and in the basis occupies just one-third part of the entire length of its diameter. Posterior to the zygomata, the petrous portions have, in the simiæ, a large development in the antero-posterior direction. Another most remarkable character, in respect to which those anatomists have been greatly deceived who compared only young troglodytes with man, is the position of the great occipital foramen—a feature most important as to the general character of structure, and to the habits of the whole being. This foramen in the human head is very near the middle of the basis of the skull, or rather it is situated immediately behind the middle transverse diameter, while, in the adult chimpantsi, it is placed in the middle of the posterior third part of the basis cranii. A third characteristic in the ape is the greater size and development of the bony palate, in consequence of which the teeth are much larger and more spread, and want that continuity which is, generally speaking, a characteristic of man; and intervals between the laniary, cutting, and bicuspid teeth, admit, as in the lower tribes of animals, the apices of teeth belonging to the opposite jaws. Fourthly, the basis of the skull is flat, owing to the want of that downward development of the brain and of the bony case, connected with the greater dimension which the cerebral organ acquires in the human being compared with the lower tribes.

The outline of the basis displays the position of the great occipital foramen, to which much importance has been affixed by anatomists in the comparison of human races. Daubenton observed that this foramen holds, in the heads of all the inferior animals, a position somewhat farther backwards than in the human head. In the human head, this foramen is near the middle of the basis of the cranium, or, as Mr. Owen has more accurately defined its position, immediately behind a transverse line dividing the basis cranii into two equal parts, or bisecting the antero-posterior diameter. In the head of the adult troglodyte the place of the foramen magnum is at the middle of the posterior third part of the basis; or, if the antero-posterior diameter is divided into three equal portions, it will be found in the midst of the third division. In the heads of young apes, which heretofore have been the subjects of comparison, this foramen is situated much more forward, or near to the middle of the basis of the skull; still its position is obviously posterior to the situation of the same foramen in the human head. Soemmering thought he perceived some difference in this respect between the skulls of Europeans and Negroes. He considered the difference to be very slight, and expressed himself with doubt upon the subject; but by all the late writers who have cited his observation, without taking much pains, as it would appear, to verify it, Soemmering's statement has been repeated in much stronger terms.

I have carefully examined the situation of the foramen magnum in many Negro skulls: in all of them its position may be accurately described as being exactly behind the transverse line bisecting the antero-posterior diameter of the basis cranii. This is precisely the place which Mr. Owen has pointed out as the general position of the occipital hole in the human skull. In those Negro skulls which have the alveolar process very protuberant, the anterior half of the line above described is lengthened in a slight degree by this circumstance. If allowance is made

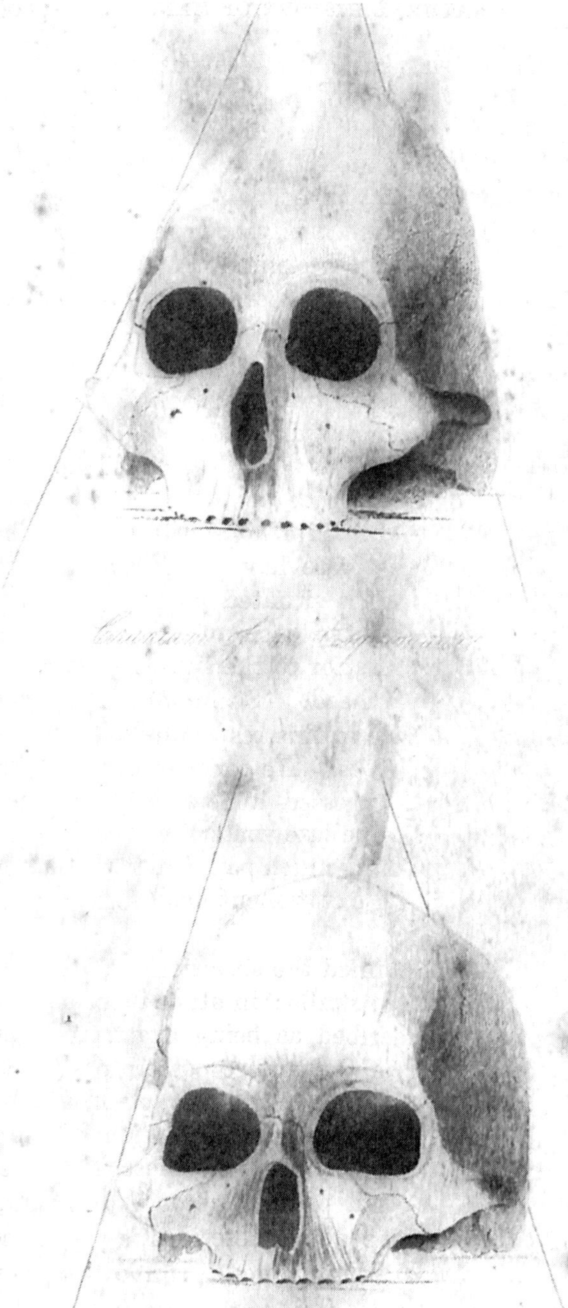

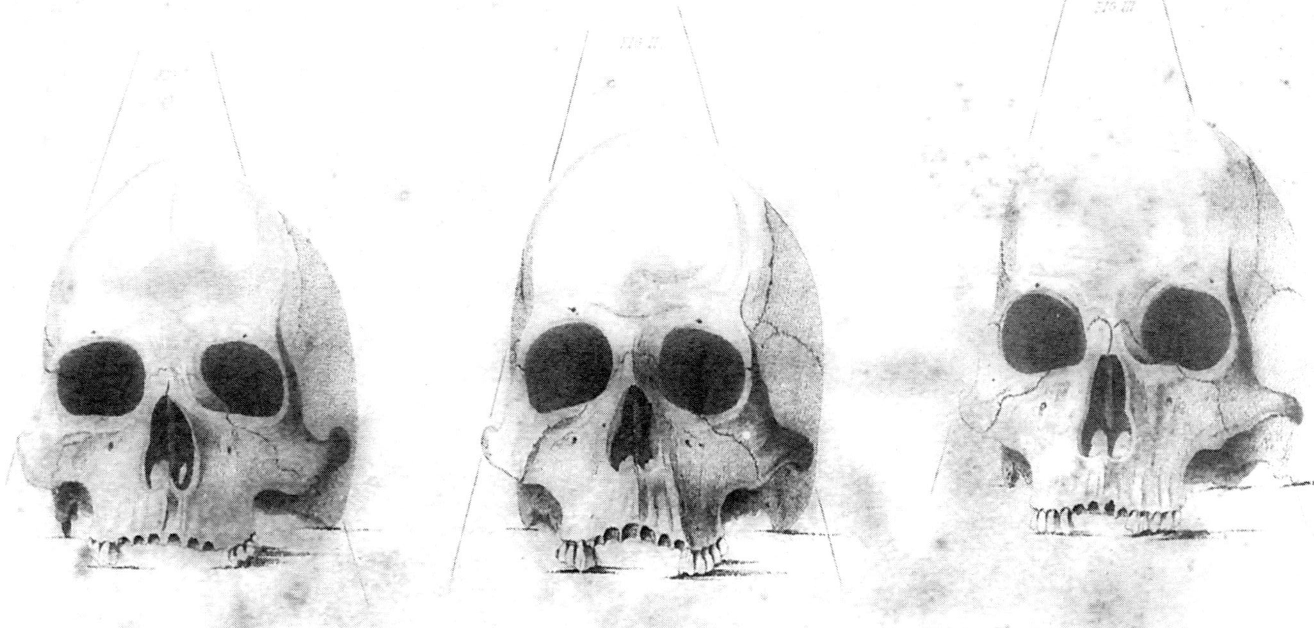

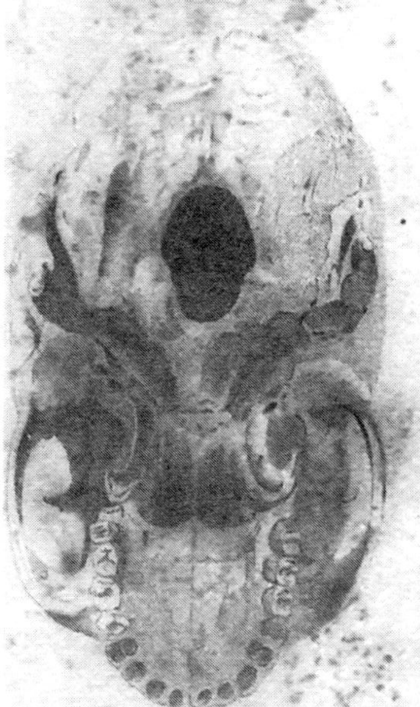

Negro, Native of Kongo

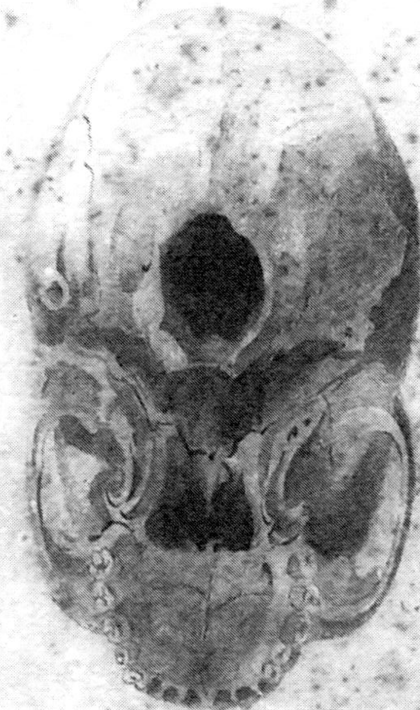

Chitimacha, Native American

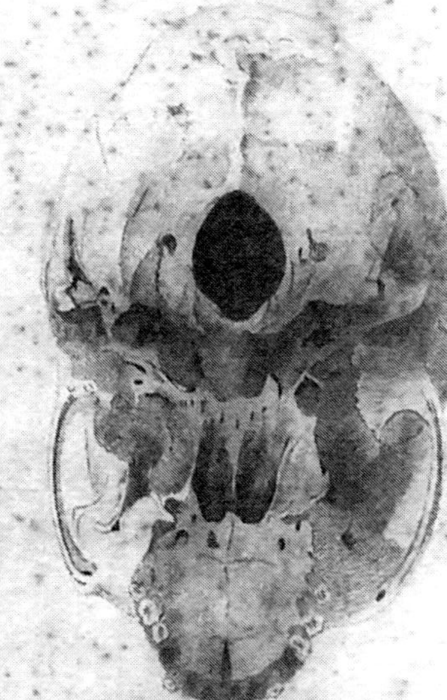

Chinese of Cant

for it, no difference is perceptible. The difference is in all instances extremely slight; and it is equally perceptible in heads belonging to other races of men, if we examine crania which have prominent upper jaws. If a line is let fall from the summit of the head at right angles with the plane of the basis, the occipital foramen will be found to be situated immediately behind it; and this is precisely the case in Negro and in European heads.

## 2. *Pyramidal Skulls.*

Neither the facial angle of Camper, nor the method of viewing the skull proposed by Blumenbach, affords a satisfactory display of the characteristics of the pyramidal or lozenge-faced skull. They are best exhibited by the front view of the face, as in Plate II. *figs.* 1 and 2, which represent strongly marked specimens of this form, and in Plate III. *figs.* 1, 2, and 3, in which it is less fully characterised. In *fig.* 1 of Plate II., which is the drawing of the skull of an Esquimaux, the lines drawn from the zygomatic arch touching the temples, meeting over the forehead, form with the basis a triangular figure. These two lines in well-formed European heads are parallel, the forehead being very much broader than in the heads of Esquimaux and other races whose skulls belong to the same great division of human crania, among whom are the Mongolians and other nomadic nations of Northern Asia. The most striking characteristic of these skulls is the great lateral or outward projection of the zygomatic arch. The cheek-bones rising from under the middle of the orbit do not project forwards and downwards under the eyes, as in the prognathous skull of the Negro, but take a direction laterally, or outwards, and turn backwards to meet a corresponding projection of the process of the temporal bone, and form with it a large rounded sweep, or segment of a circle. The orbits are large and deep. The upper part of the face being remarkably plane or flat, the nose flat, and the nasal bones, as well as the space between the eyebrows, nearly on the same plane with the cheek-bones, the triangular space described by the lines drawn on the plate may be compared to one of the faces of a pyramid. The whole face, instead of an oval form, as in most Europeans and many Africans, is of a lozenge-shape.

Another characteristic in most of the pyramidal skulls, or rather in the form of the face to which this configuration of the skull gives rise, is the apparently angular position of the aperture of the eyelids. There is no want of parallelism in the orbits, or rather of coincidence in the transverse sections of the orbital cavities. The obliquity consists in the structure of the lids themselves: the skin being tightly drawn over the large protuberance of the malar bone, under the outer angle of the eye, and at the inner extremity smoothly extended over the low nasal bones, while the bridge of the nose is scarcely elevated above the plane of the suborbital spaces, gives to the eye the appearance of being placed with the inner angle directed downwards.

The oval or elliptical form is that of Europeans, and of the Southern Asiatics who resemble them: the zygomatic bones and the jaws being in this less protuberant, the entire outline of the head, viewed from above, has no projecting angular parts, and is defined by an oval circumference. But in that oval figure, or rather ellipse, the two diameters vary considerably in proportion; in other words, some nations have rounder, others more elongated heads. The shape of the brain and of the skull at its basis is, in the rounder heads, more like that of the pyramidal skull, or the cranium of the Northern Asiatics; in the narrower heads it approaches to the figure of the elongated, or Negro head.*

* "Professor Andreas Retzius, of Stockholm, has classified the crania of the human species—firstly, according to their relative length and breadth, and secondly, according to the extent to which the parts above the jaw approach the perpendicular. By breadth, he means the diameter from

The greater relative development of the jaws and zygomatic bones, and of the bones of the face altogether, in comparison with the size of the brain, indicates, in the pyramidal and prognathous skulls, a more ample extension of the organs subservient to sensation and the animal faculties. Such a configuration is adapted, by its results, to the condition of human tribes in the nomadic state, and in that of savage hunters. Were either of these the original condition of mankind, then were the first men probably in form like the Esquimaux or the Negro. But this is an historical question with which we are not at present concerned. The physical characteristics of these last-mentioned races bear some analogy to those of the wild and uncultivated breeds of animals. But we have seen that the peculiarities of wild tribes are reproduced by a return to the wild and savage condition; and it matters not, so far as the physical process is concerned, whether the characters above alluded to in the ruder human races belonged to the primitive form, or were a subsequent result, and superinduced in nations who became barbarised, on the supposition, which is contrary to that generally entertained by naturalists, that mankind were in the beginning in a state of intellectual and moral elevation, from which they became subsequently degraded. That this last supposition is more probable than the converse one it has often been argued, without adverting to the evidence of history sacred or profane, from the consideration that rude nations, such as the Esquimaux, or the hunting tribes of Africa or America, show no tendency to civilise themselves, and that all man-

---

side to side, or parietal; by length, the diameter from the frontal to the occipital bone. The broader crania he calls brachy-cephalic; those in which the relative length predominates, dolicho-cephalic. When the parts above the jaw *retire*, the skull is prognathic; when they stand upright, *orthognathic*. The African populations are the most dolicho-cephalic, and after them, the Swedes, Germans, &c. The most brachy-cephalic are the Mongolians. Laying Africa out of the calculation, as being exclusively dolicho-cephalic, the brachy-cephalic form predominates." (*Note communicated by Dr. R. G. Latham.*)—ED.

kind, if they had been originally savage, would ever have remained without the blessings of moral and intellectual cultivation. On the other hand, there are many instances well known, and the causes are easily imagined which would lead to such a result, in which nations, after being civilised and improved, have sunk into barbarism. Whatever force there may be in this argument, there is no physical consideration opposed to it; and in comparing the different forms of the human skull, we may just as well set out from the type of the most improved races, and refer to it all the varieties which have been developed in the nomadic and savage tribes, as proceed in a contrary direction. In fact, the former method has its advantages, as it offers us at once a certain and definite model for reference and comparison.

## CHAPTER XIV.

### OF THE VARIETY IN STRUCTURE AND THE PROPORTIONS OF THE BONES DISCOVERED IN DIFFERENT RACES.

SOME varieties are well known to exist between different races of men in regard to the average stature of the body, the size and the proportions of the limbs and trunk, and the relations of different parts. These varieties have been differently estimated by anatomists; by some they have been regarded as amounting, especially when taken together with other instances of deviation, to characters truly specific, and sufficient to separate mankind into several distinct species.

Of late years, and since the attention of voyagers has been directed to the collection of facts relating to the physical history of mankind, measurements have been taken of the length of the limbs and of their proportions; and experiments have been made by means of an instrument termed a dynamometer, with the view of forming some estimate of the muscular strength in newly discovered races. The facts as yet collected are far from being so complete as to admit of a comprehensive statement; and the only general result that can be deduced is, that every tribe of people has in some of the particulars above pointed out a peculiar character. In the proportional length of limbs, in the size and relations of parts, the people of almost every island and continent have some peculiarity which might serve to distinguish them; and it would appear that in some particulars the ruder or less civilised races bear some remote resemblance to the lower animals in several particulars of their bodily conformation. These differences are discovered when we compare large numbers of one race with corresponding numbers of another. Individuals are found in every tribe who in all the particulars in question pass the intermediate line, and would be classed with the tribe which is distinguished from the majority of their own kindred. Diversities of structure, to which this observation applies, obviously cannot be regarded as specific; they are evidently but variations, since the causes which give rise to a peculiar conformation in one individual might influence a whole tribe. Races of men less improved by civilisation, like the uncultivated breeds of animals, have slender, lean, and elongated limbs. Nations who live on scanty provision of vegetable aliment are less vigorous than those who are better fed, and it would seem that the proportions of their limbs are different. The Hindús are well known to have their arms and legs longer in proportion than Europeans, and less muscular; it has been observed that when the sabres of Indian soldiers have been brought to England, the gripe has been too small for English hands. It is well known that all savage races have less muscular power than civilised men; this was first proved by the experiments of M. Péron, who found that the natives of Australia, of Timor, and of Tasmania, are feeble in comparison with Europeans; and similar trials have been made from time to time on the strength of other savage nations with like results. Mackenzie, and Lewis and Clark, assure us that the American aborigines betray the same inferiority in physical strength. In engagements between troop and troop, or man and man, the Virginians and Kentuckians, according to Volney, have always the advantage over the American savages.

In all other races compared with Europeans, the limbs are more crooked and badly formed. In the Negro, the bones of the leg are bent outwards. Soemmering and Lawrence have observed that the tibia and fibula in the Negro are more convex in front than in Europeans; the calves of the legs are very high, so as to encroach upon the hams; the feet and hands, but particularly the former, are flat, and the os calcis, instead of being arched, is continued nearly in a straight line with the other bones of the foot, which is remarkably broad.

It was observed by White, and has been generally believed, that the length of the fore-arm is so much greater in the Negro than in the European as to constitute a real approximation to the character of the ape. Facts, however, prove but a very slight difference,—one by no means greater than the varieties which are every day to be observed on comparing many individuals of any race or nation. On the other hand, the difference between adult apes and men in the length of the extremities is so great as to render all such comparisons very remote, and of very doubtful importance with respect to any ulterior conclusion. According to Mr. Owen, the arms of the orang reach to the heel, or at least to the ankle-joint; while in the chimpantsi or troglodyte they extend below the knee-joint. This is a most decided and widely marked difference between the most anthropoid apes and the uncultivated races of men. Yet even the slightest approach to the former shape would be a curious circumstance, if it could be fully established; it would tend, with other facts, to imply that the savage races of mankind have somewhat more of the animal, even in their physical conformation, than the more cultivated races, or those whose improvement by civilisation may be dated from a very remote era in the history of the world.

It has been a general opinion since the time of Soemmering that the head of the Negro is placed so much farther backward on the vertebral column as to occasion a material difference in the figure of the whole body. It was observed by Daubenton that the foramen magnum is placed, in quadrupeds, behind the centre of gravity, whence an important difference arises in the relative position of the head and trunk in man and the inferior animals. The extent of this difference, when the human skeleton is compared with that of the simiæ, has been most fully made known by Mr. Owen, who has shown that it is much greater in respect to the adult ape than it has been hitherto supposed. But there is in reality no material difference in human races; the foramen magnum is only posterior in the Negro skull to its place in the European, in consequence of the projection of the upper jaw, particularly of the alveolar process.

On surveying the facts which relate to difference in the shape of the body, and the proportions of parts in human races, we may conclude that none of these deviations amount to specific distinctions. We may rest this conclusion on two arguments: first, that none of the differences in question exceed the limits of individual variety, or are greater than the diversities found within the circle of one nation or family; secondly, the varieties of form in human races are by no means so considerable, in many points of view, as the instances of variation which are known to occur in different tribes of animals belonging to the same stock, there being scarcely one domesticated species which does not display much more considerable deviations from the typical character of the tribe. On this head enough has been said already to ensure admission of the general conclusion. More exact notice will be taken of the peculiarities of particular races when we come to describe the several tribes of people among whom they have been observed to occur as national characteristics.

# BOOK II.

## Division into Groups.

## CHAPTER I.

### OF THE DISTRIBUTION OF HUMAN RACES INTO GROUPS—AND OF THE ARRANGEMENT ADOPTED IN THE FOLLOWING SECTIONS.

We have now surveyed the extreme instances of diversity in the figure and complexion of the human races, in the view of determining whether any of them amount to specific differences; and we have shown that, separately considered, none of the physical peculiarities which distinguish the several families of men from each other exceed the limit of natural variety, or go beyond the sphere of that kind of variation which is in almost every living tribe ready to display itself under the influence of causes favourable to its development. It now remains for us to contemplate the manner in which the phenomena are by nature grouped together or distributed, and to discover the circumstances under which they display themselves.

Numerous are the divisions which different writers have adopted in distributing and classifying the varieties of the human family. Among those who consider mankind as made up of different races, no two writers are agreed as to the number of separate tribes. As there is no fixed principle of division, it seems to be in great measure arbitrary, and left to the choice of individual writers whether they shall enumerate more or fewer of such groups; and it happens that every new ethnologist subdivides the nations which his predecessor had connected, and brings together some which he had separated. As I do not follow those who have taken it for granted that mankind originally sprang from different stems, it will be superfluous for me to enter into any discussion on this head. I shall endeavour briefly to describe the principal tribes of men as I find them distinguished by historical evidence, and by that of the most authentic records, namely, by their languages, which, of all peculiar endowments, seem to be the most permanently retained, and can be shown in many cases to have survived even very considerable changes in physical and moral characters. Glottology, or the history of languages, founded on an accurate analysis of their relations, is almost a new field of inquiry. It has been explored with great success of late, and new discoveries are every day made in it. Our contemporaries are becoming more and more convinced that the history of nations, termed ethnology, must be mainly founded on the relations of their languages. The ultimate object of this investigation is not to trace the history of languages, but of the tribes of men whose affinity they tend to illustrate. We must at the same time keep in view the great physical distinctions pointed out in the preceding chapters, and particularly the three-fold divisions of the forms of the human skull. This is probably the most permanent of all physical varieties, and it must at least be taken into the account in the distribution of nations into particular departments. I shall endeavour to arrange into groups those races which appear from proofs of various kinds to have been in ancient times connected.

The most popular or generally received distribution of human races in the present day is that which was recommended by the adoption of Baron Cuvier. It did not entirely originate with that great writer, but was set forth by him in a more decided and complete manner than it had been before his time. This system refers different races of men to certain lofty mountain-chains as the local seats of their original existence. The birth-place, or the primitive station of the race of men who peopled Europe and Western Asia, is supposed to have been Mount Caucasus; from this conjecture, Europeans, and many Asiatic nations, and even some Africans, have received the new designation of Caucasians. The nations of Eastern Asia are imagined, in like manner, to originate in the neighbourhood of Mount Altai, and they are named after the Mongolians, who inhabit the highest region in that vast chain of hills. The African Negroes are derived from the southern face of the chain of Mount Atlas; they are, however, named simply the Ethiopian race, from the Ethiopians, who were the only black people known to the ancients in very remote times. A mixture of somewhat vague notions, partly connected with physical theories, and in part derived from history, or rather from mythology, has formed the groundwork of this scheme, which refers the origin of human races to high mountainous tracts. The tops of mountains first emerged above the surface of the primeval ocean, and, in the language of some philosophical theorists, first became the scene of the organising life of nature. From different mountain-tops, Wildenow, and other writers on the history of plants, derive the vegetable tribes, which they suppose to have descended from high places into the plains, and to have spread their colonies along the margins of mountain streams. High mountains thus came to be regarded as the birth-places of living races. Geological theories give their part to render these notions popular: not only the late speculations of the Count de Buffon and the learned Bailly, but the opinions of ancient philosophers, who maintained, before the time of Justin and of Pliny, that the mountains of high Asia must have been the part of the world first inhabited by men, inasmuch as that region must have been first refrigerated in the gradual cooling of the surface of our planet, and first raised sufficiently above the level of the ocean. Moreover, the poetical traditions of the ancient world describe high mountains as the scenes of the first mythical adventures of gods and men—as the resting-places on which celestial or aerial beings alighted from their cloudy habitations to take up their abode with men, and to become the patriarchs of the human race. Lofty mountains are the points in the geography of our globe on which the first dawn of historic light casts its early beams: hence the legends of the first ages begin their thread. In the cosmogony of the Hindús, it was on the summit of the sacred mountain Maha-meru, which rises in the midst of the seven dwipas, or great peninsulas, like the stalk between the expanded petals of a lotus, that Brahma, the creator, sits enthroned on a pillar of gold and gems, adored by Rishis and Gandharbhas; while the regents of the four quarters of the universe hold their stations on the four faces of the mountain. Equally famed in the ancient mythology of Iran and Zoroaster is the sacred mountain Albordj, based upon the earth, but raising through all the spheres of heaven to the region of supernal light its lofty top, the seat of Ormuzd, whence the bridge Tshinevad conducts blessed spirits of pious men to Gorodman, the solid vault of heaven, the abode of Ferours and Amshaspands. Even the prosing disciples of Confucius had their sacred mountain of Kuen-lun, where, according to the legends of their forefathers, was the abode of the early patriarchs of their race. The Arabs and the Persian Moslemin had their poetical Kâf. The lofty hills of Phrygia and of Hellas, Ida, Olympus, Pindus, were, as every one knows, famous in Grecian story. Caucasus came in for a share of the reverence paid to the high places of the earth. Caucasus, however, was not the cradle of the human race, but the dwelling-place of Prometheus, the maker of men, and the teacher of astronomy.

But all these notions are the poetical but wild dreams of men whose imaginations were excited by the splendid and unexplained meteoric phenomena of mountainous regions. It cannot be proved, nor is the supposition at all probable, that mankind began to exist till long after those physical changes had been completed which prepared our planet for the present order of the creation, and which raised a great part of the earth's surface above the level of the ocean. If we were at liberty to form a conjecture on the subject, it would be that the human race had its beginning in a comparatively late period, in a region abounding with vegetable and animal productions. There is one ancient tradition which fixes the birth-place of mankind, not on the sides or tops of snow-clad mountains, but on the banks of great rivers, which fertilise one of the most luxuriant regions of the earth. This is the tradition delivered in the Sacred Hebrew Scriptures. Hiddekel and Perath, two out of the four rivers of the Mosaic paradise, are well known and identified,—they are the Tigris and Euphrates; and at the time when the Book of Genesis was compiled, it is not to be doubted that the names of the two remaining rivers were equally well known.

I shall not attempt to trace the history of nations from the early period to which the documents of patriarchal archæology contained in the earliest portion of the Pentateuch refer. So many unfathomable chasms lie in the path, that every one of the many writers who has sought his way through the intervening wilderness has lost himself in the obscurities of doubtful speculation. Those who wish to tread on safe ground in approaching ancient times must, like the inductive philosophers, take the way *à posteriori*, and trace backwards the ever more and more evanescent vestiges of events. If in this method we endeavour to gain a distinct glimpse of the state and even of the local position of human races in the earliest periods of society, we find men collected in great numbers, not on the high and barren tracts of the earth, but on the banks and estuaries of rivers affording secure havens on the sea, and the means of communication with inland countries. The cradles or nurseries of the first nations, of those at least who became populous and have left a name celebrated in later times, appear to have been extensive plains or valleys traversed by navigable channels, and irrigated by perennial and fertilising streams. Three such regions were the scenes of the earliest civilisation of the human race, of the first foundation of cities, of the earliest political institutions, and of the invention of the arts which embellish human life. In one of these, the Semitic or Syro-Arabian nations exchanged the simple habits of wandering shepherds for the splendour and luxury of Nineveh and Babylon. In a second, the Indo-European or Japetic people brought to perfection the most elaborate of human dialects, destined to become, in after times and under different modifications, the mother tongue of the nations of Europe. In a third, the land of Ham, watered by the Nile, were invented hieroglyphical literature, and the arts in which Egypt far surpassed all the rest of the world in the earlier ages of history.

It will be found that in these three great nations, and among those who are allied to them in origin and language, are comprised nearly all the civilised communities, and, indeed, most of the tribes of people known to antiquity. When we view them as branches of the human family, it is interesting to inquire what physical differences existed among them. On this subject the ancient writers give us in general very defective information; yet they have left various notices from which we are enabled to collect proofs that the three races under consideration differed from each other in certain physical peculiarities. We find that swarthy, or brown or black people, and crisp or curly-haired tribes, with flat noses and thick lips, who had something of the African, though probably not the true Negro physiognomy, which was scarcely known to the Greeks, were, on account of their physical character alone, supposed, in the time of Herodotus, to be allied to the Egyptians. The people of Upper Asia, that is, of the Assyrian countries, are noted by Hippocrates for their great stature and the beauty of their persons, and for a certain caste of countenance in which they resembled each other more than did the people of Europe. We may consider this description as referable to the Syro-Arabian race. From various incidental descriptions of Greeks, Thracians, Italians, Celts, and Germans, we are tolerably well informed what were the physical characters of the nations of Europe. With these our third human family would coincide, were it not necessary to include among the latter some races of Southern Asia. The compound epithet of Indo-European is the best designation for this class of mankind, whom Schloezer and other German writers term Japetic, as they include the Syro-Arabian nation under the name of Semitic, or Shemites.

We cannot regard these three divisions of the ancient civilised world as representing the three great departments of mankind, as these departments are discriminated by the forms of the skull. They were neither nomades nor savages, nor do they display in their crania either of the forms principally belonging to races in those different states of existence. They had all heads of an oval or elliptico-spherical form, which we have observed to prevail chiefly among nations who have their faculties developed by civilisation. But although it cannot be said that the Egyptians had the narrow prognathous skull of the true Negro, nor the Indo-European nations the pyramidal heads of the nomadic people of High Asia, or of the Ichthyophagi of the North, yet there are not wanting marks of some relationship between the Egyptians and the nations of Central Africa, and between the Indo-Europeans and the Northern Asiatics. In their complexion, and in many of their physical peculiarities, the Egyptians were an African race. In the eastern, and even in the central parts of Africa, we shall trace the existence of various tribes in physical characters nearly resembling the Egyptians; and it would not be difficult to observe a gradual deviation among many nations of that continent from the physical type of the Egyptian to the strongly marked character of the Negro, and that without any very decided break or interruption. The Egyptian language, also, in the great leading principles of its grammatical construction, bears much greater analogy to the idioms of Africa than to those prevalent among the people of other regions. On the other hand, the languages of some of the Northern Asiatics bear tokens of relationship, though they may be admitted to be remote ones, to the idioms of the Indo-European race. The oval type of the skull prevalent among these last-mentioned nations distinguishes them, indeed, from the broad-faced Asiatics; but we can show, by many examples, that these characters are not constant, and that when nomadic nations have become settled and civilised, they have acquired a form of head similar to that of Europeans. It must be admitted that these approximations require further inquiry and more precise proofs before they can be admitted as furnishing the groundwork of an ethnological system. I shall take them as suggesting a certain arrangement or classification in the following outline, and shall thus bring together departments of mankind which it is most interesting to compare, and the comparison of which is likely to elucidate some questions in the physical history of our race. The first nation in the following series will be the Syro-Arabian, which a celebrated French anatomist, Baron Larrey, regards as the model of perfection, and at the same time the prototype of the human family. These nations hold a central place, and are cut off from contact with barbarians on either side by the nations who follow them in the series. The second group are the Egyptians, and the third the Indo-Europeans. The physical differences between all these nations, though considerable, are not greater than most persons will think sufficiently explained by reference to climate, and to diversity of food and of manners. After the Egyptians, we proceed to describe the great body of the nations of Africa, and after the Indo-Europeans, the people of High Asia. Among the former are many tribes in the lowest stages of savage life, supporting their existence on the natural fruits of the earth, or on the casual produce of the chase, and dwelling almost without houses and clothing amid the forests. The latter were chiefly nomades: the nature of the climate, and the countries which they inhabit, cold and bleak, and consisting of vast steppes, and affording no spontaneous contribution to the support of human life, precludes the existence in those regions of people reduced below the condition of wandering shepherds, possessed of some wealth, and exercising some of the simplest arts,—acquainted with the use of clothing, tents, and waggons. Men deprived by indolence or misfortune of such possessions would perish in the wilds of Tartary: on the banks of the Senegal or of the Quorra they would degenerate into the state of savages. These classes of nations have, as we have seen, different physical characters; among the African savages, we find the prognathous form of the head and all its accompaniments; and these traits display themselves in proportion to the moral and physical degradation of the race. In Northern Asia, most of the inhabitants have the pyramidal and broad-faced skulls.

After describing the nations of Africa and Northern Asia, we shall proceed to the native tribes of America, and to those of the Austral seas and the great Southern Ocean.

Having thus pointed out the order in which the remaining subjects of this work are to be arranged, I shall not advert again to this topic, but proceed successively to describe the most remarkable races of men.

## CHAPTER II.

### OF THE SYRO-ARABIAN OR SEMITIC RACE.

THE Syro-Arabian nations, termed by Eichhorn and other German writers Semitic, occupied a region of Asia intermediate between the countries of the Egyptian and the Indo-European races. They differed, as we have remarked, from both these races in their physical and moral characters. According to ancient authorities cited by Strabo, which to Bochart and Heeren appeared entirely worthy of credit, the dominion of these nations reached northwards to the Euxine, including the country of the Cappadocians, who were called by the Greeks, in the time of Herodotus, Leuco-Syri, or White Syrians. To the eastward, it bordered on Armenia and Persia, and extended southward to the Indian Ocean, comprehending the region watered by the great rivers of Mesopotamia; Syria, Palestine, and Arabia, were its western and southern parts, unless we ought to include in it some portions of Africa, which appear to have been colonised in early times by people who spoke dialects of the Syro-Arabian language.

The bond of union between all these nations, who differed widely in manners,—some being nomadic, others agricultural, and a third class devoted to foreign commerce and domestic manufactures,—is their remarkable language, so different from all other human idioms. Their language also affords the best and most applicable method of subdividing them into groups. The idioms spoken by the ancient* and modern nations of this family may be classed under the four following departments:—

1. The northern and eastern branch, termed Aramæan, or Syrian. The Syriac of the versions, and the Chaldee of the late Scriptures of the Old Testament and of the Targums, are specimens of this language from early times. If the Cappadocians were really Syrians, this was doubtless their idiom. It appears to have been the original idiom of the Hebrews, until the Abramidæ occupied the Promised Land in Canaan, and adopted, as it would appear, from its previous inhabitants, the Canaanitish, or proper Hebrew:

* The cuneiform inscriptions of Assyria and Babylonia are written in a cognate idiom, allied to, though different from, the Hebrew.—Ed.

It has been supposed by several German writers, particularly by Michaelis and Schloezer, that the Chaldees, or Chasdim, were a distinct people from the Assyrians and Syrians, and that the primitive Chaldæa was a region to the northward of Syria and Mesopotamia. The Chaldees, or rather the Chasdim, are frequently mentioned by the sacred writers towards the later periods of the royal dynasties of Judah and Samaria as a warlike people from the north. By Greek writers, the Chaldæi are connected with the Carduchi and the Chalybes, nations of barbarous mountaineers, who occupied some parts of the high region of Kúrdistan, and the former of whom first became known to Europeans by opposing the retreat of Xenophon. It is supposed by Michaelis that these Chasdim of the mountains, who are conjectured by that celebrated Oriental scholar to have been, not a Semitic but a Scythian, or perhaps a Slavonian tribe, made an inroad on the plains of Mesopotamia about the time of Isaiah, and there established a new empire, which was that of the later Chaldæan or Babylonian sovereigns. History affords very slender support of any such hypothesis, which seems to have no other groundwork than a few fanciful derivations of the names of some Babylonian sovereigns.*

2. The Hebrew, or Canaanitish, or Phœnician,—for they were the same, or very nearly the same, as Gesenius has proved,—was spoken by the Hebrews from the time when they adopted it on their arrival in Palestine to the Babylonish captivity, when they are supposed to have exchanged it for Chaldee, or to have returned to the use of a dialect more akin to their primitive ante-Abramic speech. This language, with perhaps some very slight variations, was the idiom of the Sidonian and Tyrian states, and of

* Michaelis, "Specim. Geograph. Hebræor. Ext." Part II. p. 80.—Also Schloezer, "von den Chaldäern," in Eichhorn's "Repertor. für bibl. und morgenländl. Lit." Th. 8.—Also Dr. J. R. Forster's "Epistola de Chaldæis," in Michaël. "Spec. Geog. Hebr. Ext."—See, also, Adelung's remarks on this subject, in the first part of the Mithridates.

Carthage and the Carthaginian settlements. Even the language of Numidia is supposed by Gesenius to have been a pure, or nearly pure, Hebrew. On the other hand, no traces of Phœnician inscriptions have been discovered beyond or to the northward of the Straits of Hercules;* and Gesenius denies that any proofs exist of the colonisations thought by various modern writers on history and antiquities to have been formed by Phœnicians on different points of the coast of Europe.

3. The third division of Syro-Arabian dialects are those of the Arabic properly so called, including the Moggrebyn, or Western Arabian language.

4. It is supposed that a fourth language, belonging to the Syro-Arabian stem, has been discovered lately in the southern parts of Arabia. It has been maintained by M. Fresnel that the barbarous inhabitants of Mahrah still speak the idiom that prevailed at the court of the Queen of Saba, namely, the dialect of the Hhimyarite Arabs, the Homerites of the Greeks. This idiom is termed by M. Fresnel, who has investigated its forms, Ekhkili; which is, as he says, the proper national designation of the noble race who inhabit the mountains of Hhacik, Mirbât, and Zhafar, on the southern coast of the Arabian peninsula.†

* Phœnician inscriptions have been found in Spain, and even in the south of France, at Marseilles, though it may be alleged that they were not in their original localities.—ED.

† Papers by M. Fresnel in several numbers of the "Nouveau Journal Asiatique." Paris. [A short vocabulary and some remarks on the language of the Mahrah tribes, together with some brief notices of their habits and physical features, will be found in the Journal of the Bombay Branch of the Royal Asiatic Society, vol. ii. p. 339 (1847). Assistant-Surgeon Carter, the writer of the paper alluded to, gave in the January number for 1844, p. 195, some account of the Garah tribe, who live in the immediate neighbourhood of the Mahrahs, and speak a dialect of the same variety of Arabic. Both these tribes consider themselves descendants of the Himyarites, and Edrisi says that the people of Mahrah speak the ancient language of Himyar. ("Recueil de Voyages," vol. v. p. 150.) There is, however, a very great difference in the form and stature of the tribes, the Garahs being tall, handsome, and comparatively fair, with a tinge of red; while the Mahrahs are small and ugly, with sunken eyes, and a sinister cast of countenance.

It approximates more in its forms to the Hebrew and the Syriac than to the modern or ancient Arabic,—a fact which illustrates the assertion of ancient writers, who declare that the Phœnicians came originally into Palestine from the borders of the Erythræan Sea, or the Indian Ocean. The Homerites are said to have been the Shemite people who passed the Red Sea, and founded the Abyssinian kingdom of Axoume, or Axum, where the Gheez language, or Ethiopic of the version of the Bible, and of other of the sacred books of the Abyssinian church, was the prevailing language as early as the age of Frumentius, and probably many centuries before his time. The discoveries reported by Lieutenant Wellsted and others, of inscriptions in different parts of Oman, or Southern Arabia, in characters altogether unlike the Cufic, or the oldest forms of letters known among the Northern Arabs, and bearing a striking conformity with the letters of the Gheez, give additional support to the opinions of M. Fresnel, and render it probable that an ancient language cognate with the Syriac, and the Hebrew, and Arabic, but distinct from all, and having a character of its own, once prevailed over an extensive region to the southward of that occupied by the proper Arabic.* Perhaps this was the idiom of the Cushite Arabs, who are reckoned a more ancient race than the Joktanidæ, or the tribes descended from Joktan, the

The extreme poverty of these people, who live on the coast, and get nothing to eat but fish, may account for their inferiority to the better-fed, flesh-eating Garahs who dwell further inland. The women do not veil their faces in either tribe, and both are said to be wholly without religion. It is a curious circumstance that in saluting each other the Mahrahs bring their noses in contact, inhaling the air audibly, precisely like the New Zealanders.—ED.

* These inscriptions have been published by the Geographical and Asiatic Societies, and learned investigations have been made by Gesenius, Rödiger, Dr. Wilson, Dr. Bird, and others. The general result of all that has been published is a conviction that the language is Semitic, and the alphabet the original of the Abyssinian, notwithstanding the opposite direction of the writing; but the most learned and persuasive of the investigators are the most convinced how uncertain are the interpretations they have suggested.—ED.

traditional ancestor of the Arabian tribes, and more nearly allied to the Phœnicians or Canaanites, both being Chamite, and not Shemite, nations, as we learn from the Biblical genealogies.*

Of the several nations who are connected by this community of language, some who were formerly celebrated have become nearly extinct; while others have spread themselves, either as the exiled followers of a persecuted faith, or as the conquering apostles of a victorious one, over the world, and seem destined, through the energy of their invincible mind, to survive to the end of time. The Syrian race scarcely exists: their language only survives in some districts on the borders of Kúrdistan;† everywhere else they have been lost under the predominant Arabs. The Homerites in Arabia, if there they exist, are little known: the Abyssinian Homerites are the only inhabitants of the province of Tigre, to the eastward of the Tacazze, whose idiom still resembles the ancient Gheez. The Arabs, who spread Islam by their victories from the Atlantic to the Ganges, and the Jews, who are wanderers over the whole world, are perhaps now more numerous than were even their forefathers.

The Jews have assimilated in physical characters to the nations among whom they have long resided, though still to be recognised by some minute peculiarities of physiognomy. In the northern countries of Europe they are fair, or xanthous; blue eyes and flaxen hair are seen in English Jews; and in some parts of Germany the red

* A Himyaritic inscription of five lines, apparently the commencement of a longer one, was found at Warka, in Southern Chaldæa, in the present year (1854) by Mr. Loftus; and a genuine Babylonian cylinder, with two lines in the same language and character, was deposited in the British Museum, a few months ago, by Captain Jones, who brought it from Bagdad.—Ed.

† It is spoken only in a few villages, and there in a very corrupt form, with mixture of Arabic, Turkish, and Persian. A copy of the Doctrina Christiana has been brought to Europe, and portions published in Ewald's Zeitschrift, with valuable philological notes.—Ed.

beards of the Jews are very conspicuous. The Jews of Portugal are very dark. Jews, as it is well known, have been spread from early times through many countries in the eastern parts of Asia,—in China, Tartary, and the northern parts of India. There are many of them in the towns of Cochin and the interior of Malabar. They hold communication with each other in their eastern colonies, which appear to be of one stock or migration; but at what era they reached these countries it is unknown. Their residence in Cochin appears to have been from ancient times; and they are now black, and so completely like the native inhabitants in their complexion that Dr. Claudius Buchanan says he could not always distinguish them from the Hindús.* He has surmised that the blackness of the Jews spread through different parts of India is attributable to intermarriages with Hindús; but of this there is no evidence: it is probable that the preservation of the Jews in these countries as a distinct people is owing, as elsewhere, to their avoiding all intermixture with the native inhabitants. The Jesuits in China expressly inform us that the Jews settled in Honan, where they have been established for many ages, keep themselves distinct, and intermarry within their own community.† It appears that the ancient Jewish inhabitants of Cochin were a people of the same migration with those of China; and it is very improbable that they differ from their brethren in the particular above alluded to.

The Arabs are partly shepherds, termed in their own language Ebn-el-Arab and Bedawi, whence the European name of Bedouins;—in part tillers of the earth, or Felahin; or Haddri, that is, dwellers in towns. The agricultural Arabs are of larger and stouter form; the Bedouins are thin and meagre. We are assured by Mr. Fraser that the genuine Arabs are rather spare and active than athletic

* There is at Mattacheri, a town of Cochin, a particular colony of Jews, who arrived at a later date in that country, and are called Jerusalem, or White, Jews.

† Duhalde. Astle's Voyages, vol. iv. p. 227.

men: "Those of the superior orders who came under our observation, as the Sheiks and their families, bore a strong characteristic resemblance to each other in features. The countenance was generally long and thin; the forehead moderately high, with a rounded protuberance near its top; the nose aquiline; the mouth and chin receding, giving to the line of the profile a circular rather than a straight character; the eye deep set under the brow, dark and bright. Thin and spare, deficient in muscle, their limbs were small, particularly their hands, which were sometimes even of feminine delicacy. Their beards were almost always of a deep black, artificially coloured, if not naturally so; a few wore them grizzled; and we observed an old man whose beard, of a milk-white colour, he had dyed yellow, which, contrasted with a singular pair of blue eyes, had a very extraordinary effect."

M. De Pagés has described the Arabs of the desert between Bassora and Damascus. He says, "They run with extraordinary swiftness, have large bones, a deep brown complexion, bodies of an ordinary stature, but lean, muscular, active, and vigorous. The Bedouins suffer their hair and beards to grow; and, indeed, among the Arabian tribes in general, the beard is remarkably bushy. The Arab has a large, ardent, black eye, a long face, features high and regular, and, as the result of the whole, a physiognomy peculiarly stern and severe. The tribes who inhabit the middle of the desert have locks somewhat crisped, extremely fine, and approaching the woolly hair of the Negro."

The general complexion of the Arabs on the coast of Yemen is the same yellow, bordering on brown, which is evidently the natural colour of the race, and not derived from intermixture with Africans. Niebuhr says of them, "Les femmes Arabes des contrées basses, et exposées aux chaleurs, ont naturellement la peau d'un jaune foncé, mais dans les montagnes on trouve de jolies visages même parmi les paysannes."

The skulls of the Arabian race furnish, as we have seen, in the opinion of Baron Larrey, the most perfect type of the human head. This writer observes that the skulls of the Arabs approach to a spherical figure, with a remarkable elevation of the upper part. "The heads of this race display, in other respects, the greatest physical perfection,—a most perfect development of all the internal organs, as well as of those which belong to the senses." He says that experience has proved to him that their intellectual perfectibility is proportional to this higher development of physical organisation, and that it is, *without doubt*, superior to the faculties of those nations who inhabit the northern regions of the globe, meaning the Europeans. "In Egypt," he says, "we have observed that young Arabs of both sexes imitate all the productions of our artists and artisans with astonishing facility, and that they also acquire languages with equal ease. Independently of the elevation of the vault of the cranium, and its almost spherical form, the surface of the jaws is of great extent, and on a straight perpendicular line; the orbits, likewise, are wider than they are usually seen in the crania of Europeans, and they are somewhat less inclined backwards: the alveolar arches are of moderate size, and they are well supplied with very white and regular teeth; the canines, especially, project but little. The Arabs eat little and seldom of animal food. We are also convinced that the bones of the cranium are thinner in the Arab than in other races, and more dense in proportion to their size, which is proved by their greater transparency."

In other parts of the skeleton the Arabs display, according to Baron Larrey, a proportionate superiority in organic perfection to other races of men. The following observations are important and interesting, as they relate to the history of a race which, all its branches, the Hebrew and the Phœnician being included, must be considered as the first and greatest of the whole human family. "We have observed," says M. Larrey, "first, that the convolutions of the brain, whose mass is in proportion to the cavity of the cranium, are more numerous, and the furrows which separate them are deeper, and the matter which forms the organ is more dense or firmer, than in other races; secondly, that the nervous system, proceeding from the medulla oblongata and the spinal chord, appears to be composed of nerves more dense in structure than are those of Europeans in general; thirdly, that the heart and arterial system display the most remarkable regularity and a very perfect development; fourthly, that the external senses of the Arabs are exquisitely acute and remarkably perfect: their sight is most extensive in its range; they hear at very great distances; and can, through a very extensive region, perceive the most subtile odour."

The muscular or locomotive system is strongly marked: the fibres are of a deep red colour, firm, and very elastic; hence the great agility of this people. This physical perfectibility is very far from being equalled by the mixed nations of a part of Africa and of America, and especially by the northern nations of Europe.

"Upon the whole," says Baron Larrey, "I am convinced that the cradle of the human family is to be found in the country of this race."

Such is the opinion of a very distinguished physiologist on the characters of this race, which is, as he says, confirmed by the results of long personal observations among the people of the four quarters of the globe. If the organic perfection of the Shemite nations is really, as he thinks, superior to that of other human races, to what causes are we to attribute the difference? Is the climate of Palestine and Arabia more favourable to the perfection of organisation than that of other countries? or were the higher faculties of this race more anciently cultivated than those of the Northern nations?

The complexion of the Arabs displays great diversities in the different countries inhabited by them. Volney says that some of the Bedouins are black. Niebuhr and De

Pagès assure us that the colour of the lower orders is naturally a dusky or yellow brown. According to Burckhardt, the Arabs in the low countries of the Nile bordering on Nubia are black. This traveller carefully distinguishes the Arabs from Negroes and Nubians. Higher up the Nile than Dongola are the Shegya Arabs, of whom we have an excellent description from an intelligent English traveller. "The general complexion of the Shegya Arabs," says Mr. Waddington, "is a jet black. The Shegya," he adds, "as I have already mentioned, are black,—a clear, glossy, jet black, which appeared to my then unprejudiced eyes to be the finest colour that could be selected for a human being. They are distinguished in every respect from the Negroes by the *brightness* of their colour; by their hair and the regularity of their features; by the mild and dewy lustre of their eyes; and by the softness of their touch, in which last respect they yield not to Europeans." It appears from the account given by Burckhardt and Rüppell that the Arabs on the Nile do not intermarry with the natives. The blackness of their complexion is, therefore, owing to climate alone.*

In the northern, and particularly in more elevated countries, the complexion of the Arabs is as fair as that of Europeans. Bruce says, "The Arab women are not black; there are even some exceedingly fair." He gives a remarkable description of the mountains of Ruddua, near Yambo, on the coast of Yemen, and of their inhabitants. They are high craggy mountains, abounding in springs of water and verdant spots, where various fruits grow in abundance. "The people of the place have told me that water freezes there in winter, and that there are some of the inhabitants who have red hair and blue eyes,—a thing scarcely ever to be seen but in the coldest mountains of the East."

* For a fuller account of the Shakié, or Shegya, I may refer to the second volume of my "Researches into the Physical History of Mankind," last edition. The opposite plate (Plate IV.) is a portrait of a chief, or melik, of the Shegya.

# CHAPTER III.

## OF THE EGYPTIAN RACE.

THOUGH inhabiting from immemorial times regions in juxtaposition, and almost contiguous to each other, no two races of men can be more strongly contrasted than were the ancient Egyptian and the Syro-Arabian races: one nation full of energy, of restless activity, changing many times their manner of existence,—sometimes nomadic, feeding their flocks in desert places,—now settled, and cultivating the earth, and filling their land with populous villages, and towns, and fenced cities,—then spreading themselves, impelled by the love of glory and zeal of proselytism, over distant countries; the other reposing ever in luxurious ease and wealth on the rich soil watered by their slimy river, never quitting it for a foreign clime, or displaying, unless forced, the least change in their position or habits of life. The intellectual character, the metaphysical belief, and the religious sentiments and practices of the two nations, were equally diverse; one adoring an invisible and eternal Spirit, at whose almighty word the universe started into existence, and "the morning stars sang together, and the sons of God shouted for joy;"* the other adorning splendid temples with costly magnificence, in which, with mysterious and grotesque rites, they paid a strange and portentous worship to some foul and grovelling object,—a snake, a tortoise, a crocodile, or an ape.† The destiny of the two races has been equally different: both may be said still to exist; one in their living representatives, their ever-roving, energetic descendants; the other reposing in their own land—a vast sepulchre, where the successive generations of thirty centuries, all embalmed, men, women, and children,

---

* Job, chap. xxiv.

† Clemens Alexandrin. Pædag. lib. iii. Origenes adv. Celsum, lib. iii. p. 121.

with their domestic animals, lie beneath their dry, preserving soil, expecting vainly the summons to judgment—the fated time for which is to some of them long past—before the tribunal of Sarapis or in the hall of Osymandyas. The physical characters of these nations are likewise different: instead of the sharp features, the keen, animated, and restless visages, and the lean and active figures of the Arabian, there were to be seen in the land of the Pharaohs "full, but delicate and voluptuous, forms; countenances sedate and placid; round and soft features; with eyes long, almond-shaped, half-shut, and languishing, and turned up at the outer angles, as if habitually fatigued by the light and heat of the sun; cheeks round; thick lips, full and prominent; mouths large, but cheerful and smiling; complexions dark, ruddy, and coppery; and the whole aspect displaying, as one of the most graphic delineators among modern travellers has observed, *the genuine African character*, of which the Negro is the exaggerated and extreme representation."*

There is no ancient people of whose personal characters, and manners, and habits of life, we have half so many testimonies as of those of the Egyptians, and yet there is certainly none respecting whose physical history so much difference of opinion has prevailed. Volney and many others have insisted on the assertion that the Egyptians were true Negroes. Others declare that they were a Caucasian race, or entirely of European form. Dénon, whose testimony I have just cited, has, if I am not mistaken, found the true medium. There is some excuse for this diversity of opinion in the accounts hard to reconcile which have been left us by the ancients, and the other testimonies to be collected from various quarters. I shall not put the confidence of my readers to too severe a test by requiring them to adopt my opinion in this matter, but shall hasten briefly to lay before them the principal points of evidence, from which they will form their own judg-

* Dénon, "Voy. en Egypte."

ment; and, in the first place, I shall cite some of the ancient accounts.

1. Herodotus travelled in Egypt, and was, therefore, well acquainted with the people from personal observation. He does not say anything directly as to the description of their persons, which were too well known to the Greeks to need such an account; but his indirect testimony is very strongly expressed. After mentioning a tradition that the people of Colchis were a colony from Egypt, Herodotus says that there was one fact strongly in favour of this opinion. The Colchians were μελάγχροες and οὐλότριχες,—"black in complexion, and woolly-haired." These are the words by which the complexion and hair of Negroes are described. In another passage he says that "the pigeon said to have fled to Dodona, and to have founded the oracle, was declared to be black, and that the meaning of the story was this. The oracle was in reality founded by a female captive from the Thebaid: she was black, being an Egyptian." Other Greek writers have expressed themselves in similar terms. Æschylus, in the "Supplices," mentions the crew of the Egyptian bark as seen from an eminence on the shore: the person who espies them concludes them to be Egyptians from their black complexion:—

> Πρέπουσι δ' ἄνδρες νήϊοι μελαγχίμοις
> Γυίοισι λευκῶν ἐκ πεπλωμάτων ἰδεῖν.
>
> "The sailors, too, I marked,
> Conspicuous in white robes their sable limbs."

There are other passages in ancient writers in which the Egyptians are mentioned as a swarthy people, which might, with equal propriety, be applied to a perfect black, or to a brown or dusky Nubian. We have in one of the Dialogues of Lucian a ludicrous description of a young Egyptian, who was represented as belonging to the crew of a trading vessel in the Piræus. It is said of him that, "besides being black, he had projecting lips, and was very slender in the legs, and that his hair and the curls bushed up behind marked him to be a slave."

Ammianus Marcellinus has given an account of the Egyptians, which is somewhat differently expressed:—"Ægyptii, plerique, subfusculi sunt et atrati, magisque mœstiores, gracilenti, et aridi." By saying that Egyptians, *for the most part,* are of a brownish, or somewhat brown colour, and of a tanned or blackened hue, the writer shows that this was not the case equally, at least with all of them; and the expressions *subfusculi* and *atrati* are very different from *nigri* or *atri*.

Two old Egyptian deeds of sale have come to light, which afford a very curious testimony.* Both of them belong to the Ptolemaic period; but the names of the persons mentioned indicate them to have been native Egyptians. The persons interested in the contracts are described according to their external appearance and colour. In one of these documents, the seller, who is named Paminthes, is termed μελάγχρως, and the buyer μελίχρως; which may be rendered "of a black," or, perhaps, a "dark brown colour," and yellow, or honey-coloured. The same epithet is given to the buyer, who is named Osarreres, in the other manuscript. The shape of the nose and features is also stated, but not in such terms as to give any idea of the Negro physiognomy.

From these accounts we may safely infer that the Egyptians were a dark-coloured people, and at the same time that great varieties existed among them, which is the case among the Abyssinians and the Hindús in the present day. It may be remarked that in climates like those of Europe and of Nigritia, where the influence of external agents is very strongly exerted on the races of men, their complexions display little variety. Men are either white or black according as they live under the equator,

* The facsimile of one of these documents is at Berlin; the original of the other at Paris. An interpretation of the former was given by Professor Boeckh, and that of the other by M. H. St. Martin. See Heeren, Ideen, 2, 2, Absch. 1. "Ansicht des Landes und Volkes;" and K. O. Müller, "Handbuch der Archäologie," &c. Breslau, 1830.

or at a distance from the tropics; but, in intermediate regions, both black and white individuals are found in the same nation. This is the case in India and in Abyssinia, in both of which countries the fact, as we shall see, has been commented upon by travellers who have not known how to explain it. A similar variety seems to have existed among the old Egyptians.

2. The prevailing complexion, however, of the people of Egypt has something very remarkable. If we may form an idea of the complexion of the Egyptians from the numerous paintings found in their temples, and in splendidly decorated tombs, in some of which the colours are known to be preserved in a very fresh state, we must conclude that this people were of a red copper or light chocolate colour, and that they resembled the reddest of the Fulah and Kafir tribes now existing in Africa. This colour may be seen in the numerous plates in the "Description de l'Egypte," and in the coloured figures given by Belzoni. A similar complexion is represented on the heads of the cases made of the sycamore wood, which answer the purposes of sarcophagi, and in almost all Egyptian figures. This red colour is evidently intended to represent the complexion of the people, and is not put on in the want of a lighter paint, or flesh colour; for when the limbs or bodies are represented as seen through a thin veil, the tint used resembles the complexion of Europeans. The same shade might have been generally adopted if a darker one had not been preferred, as more truly representing the national complexion of the Egyptian race. Female figures are sometimes distinguished by a yellow or tawny colour.

The features of the Egyptians are likewise represented in their paintings and sculptures, which display, in general, a very remarkable and peculiar physiognomy, or type of countenance, and bodily conformation. I have already cited the terms in which Dénon has recorded the impression produced on him by these representations.

No writer has taken greater pains in this investigation than Blumenbach, who has examined many mummies, and has in several works expressed his opinion of the physical character of the Egyptians founded on this inspection and on a study of the remains of ancient art. Blumenbach has been led to the conclusion that De Pauw, Winckelmann, and D'Hancarville, were mistaken in ascribing to the Egyptian monuments one common character of physiognomy. In Blumenbach's opinion, there are three varieties in the physiognomy expressed in paintings and sculptures, or three principal types to which individual figures, though with more or less of deviation, may be reduced: these are the Ethiopian, the Indian, and the Berberine. "The first," according to this writer, "coincides with the descriptions given of the Egyptians by the ancients: it is chiefly distinguished by prominent jaws, turgid lips, a broad flat nose, and protruding eye-balls." The second is considerably different from the first; its characters are "a long narrow nose; long and thin eyelids, which turn upwards from the bridge of the nose towards the temples; ears placed high on the head; a short and thin bodily structure, and very long shanks." As a specimen of this form, he mentions the painted female figure on the back of the sarcophagus of Captain Lethieullier's mummy, which he considers as decidedly resembling the Hindús. "The third sort of Egyptian figures partakes something of both the former. It is characterised by a peculiar turgid habit, flabby cheeks, a short chin, large prominent eyes, and a plump form of body." This is the type most generally followed in Egyptian paintings. It is supposed to represent the ordinary form of the Egyptians, and what may be termed their peculiar national physiognomy. It is thought by Blumenbach to approach very nearly to the form of the Barábra or Berberines.

The following figures afford some specimens of the characters exhibited by Egyptian sculptures. The first figure is the head of a statue supposed to be that of

Rameses. It is thought by Mr. Martin* to resemble the second Egyptian type described by Blumenbach, namely, that which approaches the Hindu.

FIG. 44.

Head of Rameses.

In this figure, it is observed that "the general expression is calm and dignified; the forehead is somewhat flat; the eyes are widely separated from each other; the nose is elevated, but with spreading nostrils; the ears are high; the lips large, broad, and turned out, with sharp edges; in which points there is a deviation from the European countenance."†

The following figures, painted in fresco, in which the

* See Mr. Martin's "Natural History of Mammiferous Animals," &c. 8vo. Plates. London, 1841.

† Ibid.

complexion is of the Egyptian red, display a physiognomy which is certainly not European.

Fig. 45.

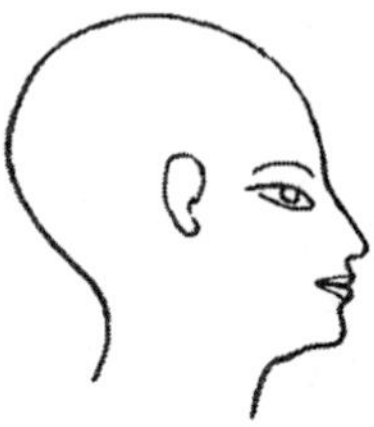

Fig. 46.

Fig. 47.

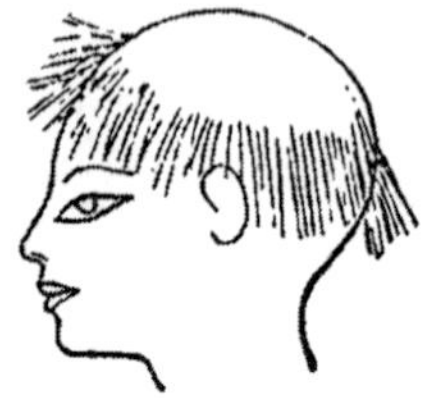

3. The Copts are well known to be the descendants of the ancient Egyptians. Egypt received a considerable number of Greek and Roman colonists; but the European settlers were probably confined mostly to the Delta and a few Grecian and Roman cities. That the Egyptian race remained nearly unaltered in the interior and remote parts of the country may be inferred from the preservation of their language, which was extant in its three dialects, with a slight admixture of Greek words, until the era of the conquest of Egypt by the Moslemin; and subsequently to that event the Christian population has been preserved by obvious causes from intermixture with strangers. Among the modern Copts many travellers have remarked a certain approximation to the Negro. Volney says that they have a yellowish dusky complexion, neither resembling the Grecian nor Arabian. He adds, "that they have a puffed visage, swollen eyes, flat nose, and thick lips, and bear much resemblance to Mulattoes." Very similar is Baron Larrey's description of the Copts, the principal traits of which are a full countenance, a long aperture of the eyelids—"coupés en amande," projecting cheek-bones, dilated nostrils, thick lips, and hair and beard black and crisp. M. Pugnet, an intelligent physician, and an ingenious and discriminating writer, has made an attempt to distinguish the Copts, or Qoubts, as he terms them, into two divisions; those whose ancestry have been intermixed, and partly of Greek or Roman descent, and a class of

purely Egyptian origin. He says that nothing is more striking than the contrast between the small and meagre Arabs and the large and fine stature of the Qoubts. "A l'extérieur chétif et misérable des premiers ceux-ci opposent un air de majesté et de puissance; à la rudesse de leurs traits, une affabilité soutenue; à leur abord inquiet et soucieux, une figure très épanouie." This description applies to both classes of the Coptic race; the following to those who are supposed to be the unmixed descendants of the old Pharaonic Egyptians. "Les Egyptiens sont en général d'une taille au dessus de la moyenne, leurs formes se prononcent vigoreusement, la couleur de leur peau est d'un rouge obscur; ils ont le front large; le menton arrondi; les joues médiocrement pleines; le nez droit; les ailes nasales fortement sinueuses; les yeux grands et bruns; la bouche peu fendue; les lèvres grosses; les dents blanches; les oreilles hautes et très détachées; enfin, les sourcils et la barbe extrêmement noirs."

M. Dénon says he was struck with the resemblance of the Copts to the old Egyptian sculptures, characterised by "*flat foreheads;* eyes half closed, and raised up at the angles; high cheek-bones; a broad, flat nose, very short; a large flattened mouth, placed at a considerable distance from the nose; thick lips; little beard; a shapeless body; crooked legs, without any expression in the contour; and long flat feet."

Mr. Ledyard, whose testimony is of the more value as he had no theory to support, says, "I suspect the Copts to have been the origin of the Negro race; the nose and lips correspond with those of the Negro. The hair, wherever I can see it among the people here (the Copts), is curled, not like that of the Negroes, but like the Mulattoes."

4. Of the skulls found in mummies. The Egyptian skulls were, as we have before observed, generally of that form which belongs to all the most anciently civilised races,

namely, the oval; but there were great varieties among the people in this respect. Most of them resembled the European skull in many particulars; but in some a certain approximation to the African has been observed or fancied. The figure in the margin is a sketch of a skull in the Museum of the College of Surgeons.

FIG. 48.

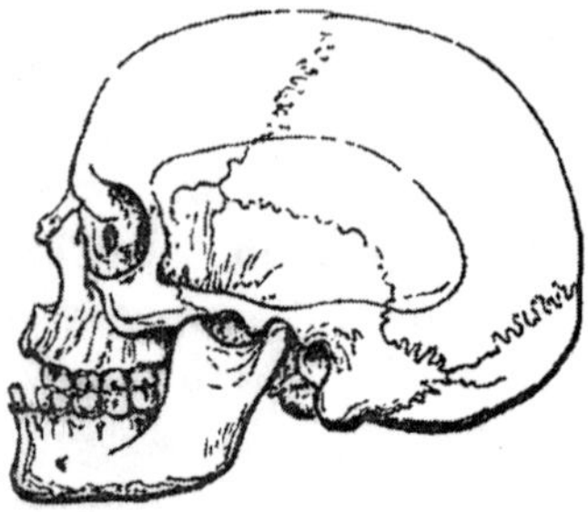

Egyptian Skull from Memphis.

There is an Egyptian skull in the same museum, which, in weight and density, resembles the heavy skulls of some Guinea Negroes. Its form is European, except that the alveolar edge of the upper jaw is rather more prominent than usual. This, with a corresponding structure of the soft parts, might have given to the countenance much of the Negro character. Soemmering has described the heads of four mummies examined by him; two of them differed in no respect from European skulls; the third, as he says, represented the African form, in having the space marked out by the insertion of the temporal muscle more extensive than in European heads. Blumenbach has published engravings of three Egyptian skulls in his "Decades Craniorum." One of these differs, as he observes, widely from the skulls of Negroes of Guinea, but has something of the Ethiopian character, and resembles the portrait of Abbas Gregorius. Another so nearly resembles the cranium of an Indian from Bengal that no material difference can be perceived between them.

## *Concluding Remarks.*

It appears, from all the evidence we have been able to collect in relation to this subject, that although the general shape of the skull among the ancient Egyptians was the oval and fully developed form which is common to highly cultivated nations, yet there were in other respects, in the physical type of that race, many tokens of relationship to the people of Africa. The puffed and full countenance, the full cheeks, thick turned-out lips, the peculiar shape of the mouth and eyes, the coppery and dusky complexion, approaching in individuals to black, in others to red, like the colour of the Fúlahs, and only a few shades lighter than that of the Berberines, who will be described in the sequel, are instances of this resemblance. The weight and density of some Egyptian skulls and the projection of the alveolar process already noticed, and the peculiar shape of the legs and flattened feet, must be taken into account. In estimating the whole amount of evidence indicative of African relations, we must take into our view many circumstances connected with the moral habits, the singular superstitions, and the general laws governing the structure of language common to the Egyptians and many other nations of the same continent. On these subjects it is impossible to enter into details in the present work, and I must beg leave to refer my readers who think it worth while to give them further consideration, to the second volume of my "Researches into the Physical History of Mankind," which is entirely devoted to African ethnography.

If it be admitted that the Egyptians display some traces of approximation in physical character to the other nations of Africa,—a fact which was striking to Ledyard and to Dénon; and if it be supposed that these traits are the result of physical agencies on a race subjected during thousands of years to their influence, it may be supposed, with great probability, that similar causes operating upon tribes of people in the rudest condition of existence, and so much the more subjected to the influence of climate and to other agencies which modify the moral and physical character of human races, would produce a much greater and more general effect. These remarks are, however, only offered as conjectures.

## CHAPTER IV.

### OF THE ANCIENT EGYPTIANS.

A MOST interesting and really important addition has been made to our knowledge of the physical character of the ancient Egyptians since the preceding chapter was written. This has been derived from a quarter where local probabilities would least of all have induced us to look for it. In France, where so many scientific men have been devoted, ever since the conquest of Egypt by Napoleon, for a long time under the patronage of government, to researches into this subject; in England, possessed of the immense advantage of wealth and commercial resources; in the academies of Italy and Germany, where the arts of Egypt have been studied in national museums, scarcely anything has been done since the time of Blumenbach to elucidate the physical history of the ancient Egyptian race. In none of these countries have any extensive collections been formed of the materials and resources which alone can afford a secure foundation for such attempts. It is in the United States of America that a remarkable advancement of this part of physical science has been at length achieved. "The Transactions of the American Philosophical Society" contain a memoir by Dr. Morton, of Philadelphia,* in which that able and zealous writer, already distinguished by his admirable researches into the physical characters of the native American races, has brought forward a great mass of new information on the ancient Egyptians. The author informs us that his present facilities have "been almost exclusively derived, directly or indirectly, from the scientific zeal and personal friendship of George R. Gliddon, Esq. late consul from the United

* Transactions of the American Philosophical Society, held at Philadelphia, for promoting Useful Knowledge, vol. ix. New Series, Part I. Article 3, "Observations on Egyptian Ethnology, derived from Anatomy, History, and the Monuments." by Samuel G. Morton, M.D.

States, in the city of Cairo." He adds, "During a former visit to the United States, this gentleman entered warmly into my views and wishes, and on his return to the East, in 1838, he commenced his researches on my behalf; and in the course of his various travels in Egypt and in Nubia, as far as the second cataract, he procured one hundred and thirty-seven human crania, of which one hundred pertain to the ancient inhabitants of Egypt. They are arranged by the latter gentleman into two series, the Pharaonic and the Ptolemaic; but, without availing myself of this classification, I have merely to regard them in reference to their national characters.

"Mr. Gliddon's residence for the greater part of twenty-three years in Egypt, and his varied official and other avocations, together with his acquaintance with the people and their languages, have given him unusual facilities for collecting the requisite materials, while their authenticity is amply vouched for by one who blends the character of a gentleman with the attainments of a scholar.

"The object of this memoir, therefore, will be to throw some additional light on the questions to which I have adverted, and to ascertain, if possible, the ethnographic characters of the primitive Egyptians; or, in other words, to point out their relative position among the races of men.

"It is necessary, however, to premise, that the materials in my possession were collected without the slightest bias of opinion on the part of Mr. Gliddon, who, at the period in question, had paid no particular attention to ethnography; and, indeed, many of these crania were received by me in their original wrappings, which were first removed, after a lapse of ages, by my own hands.

"It is further requisite to bear in mind, that, with a few exceptions, I have no clue whatever whereby to ascertain, or even to conjecture, the epoch to which these remains have belonged. The Egyptian catacombs do not always contain their original occupants, for these were often displaced, and the tombs resold for mercenary purposes;

whence it happens, that mummies of the Greek and Roman epochs have been found in those more ancient receptacles which had received the bodies of Egyptian citizens of a far earlier date. The bodies thus displaced, however, *were not destroyed;* and the Egyptians of at least twenty-five centuries before our era, though, for the most part, mingled without regard to rank or epoch, are still preserved in their interminable cemeteries."

It must be remembered that Blumenbach, whose opportunity of examining Egyptian skulls had been generally more limited than those of which Dr. Morton has so successfully availed himself, had recognised among the ancient Egyptians three craniological types, or different national shapes of the skull. These were termed by him the Ethiopian form, by which he meant the Negro or something approaching to the Negro shape; the "Hinduartige," or a figure resembling the Hindú, or that which Blumenbach supposed to be characteristic of the Hindú, a form with elongated eyes and eye-lashes, long and slender limbs, and such as is often seen on mummy-cases and in paintings; thirdly, a conformation which Blumenbach regarded as more peculiarly national and Egyptian, though he has denominated it Berber-ähnliche, or Berberin-ähnliche, from an opinion that it belonged to the inhabitants of the valley of the Nile in general, and is preserved by the modern Berberines or Barabra, as the Nubians who border on the Nile are generally called. Later researches have elucidated the history of this last type of the human cranium. In the second volume of my "Researches into the History of Mankind," I have endeavoured to collect facts which throw light upon its history, and establish the conclusion that it exists as a national peculiarity to many races of Eastern Africa. It appears, in reality, to be the form not only of the Nubian Berberines, but of many other races of Eastern Africa, as the Amhara and Galla races. It seems to coincide nearly with the native Egyptian type of Dr. Morton's classification, which is founded on the actual examination of skulls; whereas the recognition of this physical character, proper to the African races above mentioned, has been derived from the observations of travellers,* and the descriptions and delineations which they have given of different tribes.

I must give Dr. Morton's classification of the several physical characters recognised among Egyptian skulls, in his own manner of describing and arranging them.

### "*Egyptian Ethnography.*

"It was remarked, fifty years ago, by the learned Professor Blumenbach, that a principal requisite for an inquiry such as we now propose would be 'a very careful technical examination of the skulls of mummies hitherto met with, together with an accurate comparison of these skulls with the monuments.' This is precisely the design I have in view in the following memoir, which I therefore commence by an analysis of the characters of all the crania now in my possession. These may be referred to two of the great races of men, the Caucasian and the Negro, although there is a remarkable disparity in the number of each. Caucasian heads also vary so much among themselves as to present several different types of this race, which may, perhaps, be appropriately grouped under the following designations:—

### "*Caucasian Race.*

"1. *The Pelasgic Type.*†—In this division I place those heads which present the finest conformation, as seen in the Caucasian nations of Western Asia and Middle and Southern Europe. The Pelasgic lineaments are familiar to us in the beautiful models of Grecian art, which are

* Burckhardt, Rüppell, Dénon, and Larrey, and other members of the Egyptian Institute.

† "I do not use this term with ethnographic precision, but merely to indicate the most perfect type of cranio-facial outline."

remarkable for the volume of the head in comparison with that of the face, the large facial angle, and the symmetry and delicacy of the whole osteological structure. Plate III. *fig*.6,* and Plate X. *fig*. 8,* are among the many examples of this conformation.

"2. The *Semitic Type*, as seen in the Hebrew communities, is marked by a comparatively receding forehead; long, arched, and very prominent nose; a marked distance between the eyes; a low, heavy, broad and strong, and often harsh development of the whole facial structure.—Plate XI. *fig*. 2.*

"3. The *Egyptian* form differs from the Pelasgic in having a narrower and more receding forehead, while the face being more prominent, the facial angle is consequently less. The nose is straight or aquiline, the face angular, the features often sharp, and the hair uniformly long, soft, and curling. In this series of crania I include many of which the conformation is not appreciably different from that of the Arab and Hindú; but I have not, as a rule, attempted to note these distinctions, though they are so marked as to have induced me, in the early stage of the investigation, and for reasons which will appear in the sequel, to group them, together with the proper Egyptian form, under the provisional name of *Austral-Egyptian* crania. I now, however, propose to restrict the latter term to those Caucasian communities which inhabited the Nilotic valley *above* Egypt. Among the Caucasian crania are some which appear to blend the Egyptian and Pelasgic characters: these might be called the *Egypto-Pelasgic* heads; but without making use of this term, except in a very few instances by way of illustration, I have thought best to transfer these examples from the Pelasgic group to the Egyptian, inasmuch as they so far conform to the latter series as to be identified without difficulty. For examples of this mixed form, I refer especially to Plate XI. *fig*. 1,* and Plate III. *fig*. 7.*

* These refer to the plates in Dr. Morton's paper.

"*Negro Race.*

"The true *Negro* conformation requires no comment; but it is necessary to observe, that a practised eye readily detects a few heads with decidedly mixed characters, in which those of the Negro predominate. For these I propose the name of *Negroid* crania; for while the osteological development is more or less that of the Negro, the hair is long, but sometimes harsh, thus indicating that combination of features which is familiar in the Mulatto grades of the present day. It is proper, however, to remark, in relation to the whole series of crania, that while the greater part is readily referable to some one of the above subdivisions, there remain other examples in which the Caucasian traits predominate, but are partially blended with those of the Negro, which last modify both the structure and expression of the head and face.

"We proceed, in the next place, to analyse these crania individually, arranging them, for the purpose of convenience, into seven series, according to their sepulchral localities, beginning with the Necropolis of Memphis in the north:—

"First series, from the Memphite Necropolis:

"A. Pyramid of five steps.
"B. Saccára, generally.
"C. Front of the Brick Pyramid of Dashour.
"D. North-west of the Pyramid of five steps.
"E. Toora, on the Nile.

"Second series, from the Grottoes of Maabdeh.
"Third series, from Abydos.
"Fourth series, from the Catacombs of Thebes.
"Fifth series, from Kouru Ombos.
"Sixth series, from the Island of Beggeh, near Philæ.
"Seventh series, from Dehòd in Nubia."

In one very important particular, the researches of Dr. Morton have confirmed the observations of Blumenbach. It was a remark of that great physiologist, that a certain proportion of Egyptian mummies display a decided approximation to the Negro type. This is the character of the crania which Dr Morton terms Negroid; they are not genuine Negro skulls, but make an advance towards that character. Five skulls of this division were found in the Theban catacombs, and one at Maabdeh in the middle part of Egypt, and within the nome of Heracleopolis. The comparison of numbers in a synoptical table shows that eight-tenths of the whole number of crania in Dr. Morton's collection belong to the three forms which he terms Pelasgic, Semitic, and Egyptian. In these the Pelasgic form is as one to one and a half, and the Semitic as one to eight compared with the Egyptian. The *Negroid* conformation, so termed, preponderates in eight instances, thus constituting about one-thirteenth part of the whole, and the series contains no single genuine Negro. This was taken, it may be observed, from a mummy found at Philæ, on the borders of Nubia.

It must be noted that Dr. Morton ascribes the phenomena indicative of approximation to the Negro character, to mixture of races. He has not, however, recorded any fact connected with the history of the mummies in which these phenomena display themselves, or to the situation in which they are found, tending to confirm this opinion. Neither are the forms, which are termed in Dr. Morton's work "Pelasgic and Semitic," proved, by any peculiarity in the mode of sepulture or the accompanying works of art, to belong to any other than genuine Egyptians. It may be further observed, that the Copts, descended from the Egyptian race, display corresponding variations in their persons; and that the descriptions of the physical characters of the old Egyptians, drawn by ancient writers, and found in the ancient deeds before cited, impress us with the opinion that they were a people, like the Abyssinians, of very diversified characters.

Some difference of opinion may exist as to the propriety of some of the terms adopted by Dr. Morton, but he says that the first designation, "Pelasgic, is not to be considered as used with ethnological precision," or exactly in an historical sense, but merely to indicate the most perfect type of the human cranium. It cannot be maintained that the Shemite nations are at all inferior to the Europeans in the development of the brain, or of its containing bony case.*

By the first and second of these divisions we are to understand the most perfect European development, and one that is less eumorphous or less perfect, and with less ample development of the anterior part of the head, but still the Iranian, or what Blumenbach called the Caucasian, type.

We are not to suppose that these different forms, contrasted with what is termed Egyptian, are meant to be referred to races foreign to the Egyptian. This seems not to be Dr. Morton's opinion, from the following very judicious remarks, which refer particularly to the skulls found in the Memphite Necropolis.

"A mere glance at this group of skulls will satisfy any one accustomed to comparisons of this kind, that most of these possess the Caucasian type in a most striking and unequivocal manner, whether we regard their form, size, or facial angle. It is in fact questionable, whether a greater proportion of beautifully moulded heads would be found among an equal number of individuals taken at random from any existing European nation. The entire series consists of sixteen examples of the Pelasgic, and seven of the Egyptian form, a single Semitic head, one of the Negroid variety, and one of mixed conformation. Of the antiquity of these remains there can be no question, and with respect to a part of them, those from the Pyramid of five steps, we have evidence of a more precise character. It may,

* Neither of the two specimens which Blumenbach gives (as the finest examples from his collection) of the capacious intellectual head, with ample expansion of the forehead and brain, is of the Indo-European race, one being a Georgian—that is, of a really Caucasian, not European tribe—and the other, a Jewish cranium.

perhaps, be conjectured by some, that the Pelasgic heads of this series belong to the Ptolemaic epoch, and hence pertain to the Greek inhabitants of that age. But it must be remembered that the rule of the Ptolemies lasted but about 300 years, whereas the Egyptians were themselves the masters of Memphis, and entombed their dead in its necropolis, more than 2000 years before either the Persians or Greeks effected the conquest of the country, no less than during the period of and after these epochs of foreign domination.

"Of the sixteen adult Pelasgic skulls of this series two or three are small; yet the whole number gives about eighty-eight cubic inches for the average internal capacity of the cranium, or size of the brain, while the mean of the facial angle is 80°. The seven Egyptian crania have a mean internal capacity of eighty cubic inches, and a facial angle of 77°."

The following observations of Dr. Morton on the more prevalent Egyptian physiognomy are well deserving of attention. It appears that this type pervaded all ranks of the Egyptian community.

"The monuments from Meroë to Memphis present a pervading type of physiognomy, which is everywhere distinguished at a glance from the varied forms which not unfrequently attend it, and which possess so much nationality, both in outline and expression, as to give it the highest importance in Nilotic ethnography. We may repeat that it consists in an upward elongation of the head, with a receding forehead, delicate features, but rather sharp and prominent face, in which a long and straight or gently aquiline nose forms a principal feature. The eye is sometimes oblique, the chin short and retracted, the lips rather tumid, and the hair, whenever it is represented, long and flowing.

"This style of features pertains to every class,—kings, priests, and people,—and can be readily traced through every period of monumental decoration, from the early

Pharaohs down to the Greek and Roman dynasties. Among the most ancient, and at the same time most characteristic, examples, are the heads of Amenoph the Second, and his mother, as represented in a tomb at Thebes, which dates in Rosellini's Chronology 1727 years before our era. In these effigies all the features are strictly Egyptian, and how strikingly do they correspond with those of many of the embalmed heads from the Theban catacombs.

"A similar physiognomy preponderates among the royal Egyptian personages of every epoch, as will be manifest to any one who will turn over the pages of Champollion and Rosellini. The head of Horus (Plate XIV. *fig.* 2*) is an admirable illustration, while, in the portraits of Rameses IV. and Rameses VII. (Plate XIV. *figs.* 6 and 7*) the same lines are apparent, though much less strongly marked. How admirably, also, are they seen in the subjoined juvenile head (*fig.* 49), which is that of a royal prince, copied from the very ancient paintings in the tomb of Pehrai at Eletheias; and also the face of Rameses VII. (*fig.* 50), who lived, perhaps, 1000 years later in time.

FIG. 49. FIG. 50.

"I observe that the priests almost invariably present this

* These refer to plates in Dr. Morton's paper.

physiognomy; and, in accordance with the usage of thei caste, have the head closely shaven. When coloured the are red, like the other Egyptians.

FIG. 51.

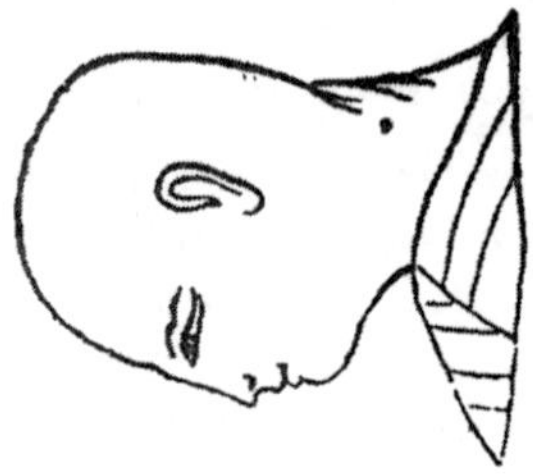

FIG. 53.

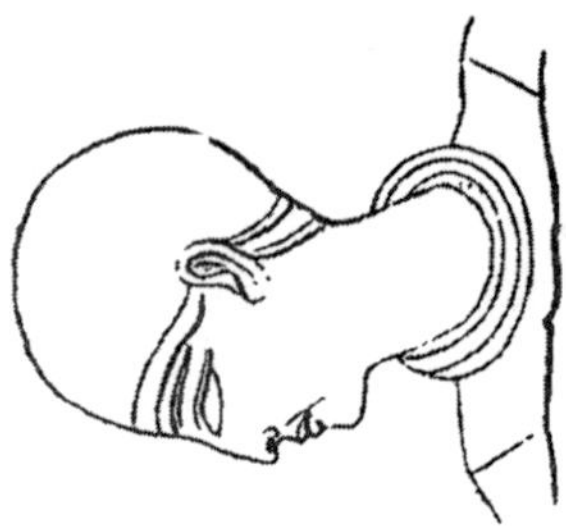

FIG. 52.

FIG. 51.

"The subjoined drawin (*fig.* 51), which is somewha harsh in outline, is from the por tico of one of the pyramids o Meroë, and is probably one o the oldest human effigies in Nu bia. They abound in all th temples of that country, an especially at Semneh, Dakkeh Toleb, Gebel-Berkel, and Mes soura.

"From the numberless ex amples of similar conformation I select another of a priest fron the bas-relief at Thebes, whicl is remarkable for delicacy o outline and pleasing serenit of expression (*fig.* 52).

"So invariably are these cha racters allotted to the sacer dotal caste, that we readil detect them in the two priest who, by some unexplained con tingency, became kings in th twentieth dynasty. Their name read Amensi-Krai-Pehor an Phisham on the monuments and the accompanying outlin is a fac-simile of Rosellini' portrait of the latter personage who lived about 1100 year before the Christian era. In this head the Egyptian an Pelasgic characters appear to be blended, but the forme preponderates (*fig.* 53).

"The last outline (*fig.* 54) represents a modification of the same type, that of the *Harper* in Bruce's tomb at Thebes. The beautiful form of the head, and the intellectual character of the face, may be compared with similar efforts of Grecian art. It dates from Rameses the Fourth.

"The peculiar head-dress of the Egyptians often greatly modifies, and, in some degree, conceals their characteristic features; and may, at first sight, lead to the impression that the priests possessed a physiognomy of a distinct or peculiar kind. Such, however, was not the case, as a little observation will prove.

Fig. 55.

"In addition to the copious remarks already made in reference to the hair, we cannot omit the annexed picture from a tomb in Thebes, which represents an Egyptian woman in the act of lamentation before the embalmed body of a relative, whilst the long black hair reaches even below the waist.

"Entering Africa by the Isthmus of Suez, the children of Ham were ushered into the fertile valley of the Nile, a region prepared by nature for settled communities and a primæval civilisation. In a country bounded by the Red Sea on one side, and by a wilderness on the other, and presenting but a narrow strip of land for its inhabitants, laws would at once become necessary for mutual protection; and we may suppose that, while one portion of the Misraimites embraced these social restrictions, another, impatient of control, passed beyond the desert barrier on the west, and spreading themselves over the north of Africa, became those nomadic tribes to which the earliest annals give the name of Libyans. It follows from this view of the question, that we suppose the Egyptians and Libyans to have been a cognate people; that the former were the aboriginal inhabitants of the valley of the Nile, and that their institutions, however modified by intrusive nations in after times, were the offspring of their own minds."

We now proceed to describe the third of the three cultivated nations of antiquity, namely, the Indo-European or Arian race. We shall first survey the nations of this family in Asia, and afterwards its colonies in Europe.

## CHAPTER V.

### OF THE ARIAN RACE.

NATIONS who speak languages of cognate origin, and who are proved by that connecting bond to be the descendants of one original stock, are spread, as it is well known, from the mouth of the Ganges to the British islands and the northern extremities of Scandinavia. They are termed collectively Indo-European nations. I shall not enter at once on so wide a field as the description of all these tribes, but shall confine myself at present to that great branch of the stock which has peopled some of the finest countries in Asia.

This great Asiatic branch, which by itself is sufficiently extensive to be termed a family of nations, and may for the present be considered as such, is divided into two principal stems. Tradition falls short of the era of separation, and we cannot trace them to one centre; but we find them both arising in the earliest periods, and in the very infancy of nations, from two principal foci at no great distance from each other, and situated to the eastward and westward of the river Indus. It is worth while to observe that they have both one common name. Arians, or Aryas,* is the ancient national designation both of the

* "Commentaire sur le Yaçna," par M. Eug. Burnouf. 4to. Paris. Annotations. Ritter, "Erdkunde von Asien," "Iranische Welt."

Persian and Indian branch. The ancient Medes called themselves Arii,* a name which survived in the Aria and Ariana of the Greek geographers.† Aryavarta was the Holy Land of the Brahmans, the country lying between the Himálaya and the Vindhya mountains, which was the ancient abode of the Hindús. In the north-western part of that region, in countries watered by the Saraswatí, the earliest traditions of the Brahmans place the ancestors of the Indian race;‡ and "Saraswatí bála báni," or "the language of children on the banks of the Saraswatí," is the distinguishing term for the Prakrit§ dialect, a vernacular or spoken language, and the oldest popular modification of the written and elaborate Sanskrit. There the Hindús had the seat of their early national existence five-and-twenty centuries before the Christian era, and thence they appear gradually to have spread, under the hierarchy of the Brahmans and their two royal dynasties, descendants of the Sun and Moon, over the different provinces of Rajputána, Ayodhya, Saurashtra, and farther eastward to Indraprest'ha, or Delhi, and to Magadha and the Gangetic provinces. To the northward they advanced into the celebrated valley of Kashmír, the basin of an ancient lake, which the Saint or Muni Kasyapa laid bare by cleaving with his scymetar its rocky sides. The aboriginal mountaineers of the Himálaya were foreign to the Indian race; but it would seem that an offset of the same stock was very early planted on the heights of the Hindu-Kúsh, or Indian Caucasus, and on the border of the cold and lofty region of Pamer, where they have dwelt unknown from remote times, designated as Kafirs by surrounding

* Herod. lib. vii. c. 62.

† Strabo, "Geog." p. 724. Ed. Casaub.

‡ "Institutes of Menu," book ii. 17, 18. Wilson, "Preface to Vishnu Purana." Elphinstone's "Hist. of India," vol. i. p. 388.

§ Colebrooke's "Essay on the Sanskrit and Prakrit languages." [Colebrooke gives the above as his interpretation, but says that the commonly received meaning is, "youthful speech of Saraswati." See his Miscellaneous Essays, vol. ii. p. 21.—ED.]

Mohamedans. They retain very few vestiges of their Indian original, except their Sanskrit speech;* while in their sanguine complexion and xanthous hair they resemble the fairest of the Danes or Swedes. The natives of the Dekhan, cut off from Hindústan by the Vindhya chain, are a distinct race, and speak dialects not of the Sanskrit but of the Tamulian family of languages. Even in Aryavarta itself it has often been conjectured that a part of the inhabitants are relics of an ancient population conquered by the Brahmans. This may be true with respect to the Pariahs, or outcasts, but it cannot be supposed with probability that the Sudras, who are reckoned among the offspring of Brahma, though they issued from his feet, were of an entirely foreign stem; still less can it apply to either of the three twice-born classes which included the Brahmans, the Kshatriyas, and the Vaisyas, likewise termed Aryas, who formed the great body of the Indian nation.

To the westward of the Indus, not far from Bamian, or from Balkh, in the ancient Bactria,† according to Lassen and Burnouf, who have for the first time elicited an historical sense from the fragments of the Magian scriptures in the Vendidad and the Bundehesh,‡ was the country which the earliest traditions of the Persians point out as the primeval seat and paradise of their race. "Eeriene Veedjo, or the pure Iran, was the region of all delights, till Ahriman, the evil one, made in the river which watered Eeriene the serpent of winter."§ The people of Ormuzd

* "Einige ethnographische Mittheilungen über die Siah-Pôsh," von Alex. Burnes, in the Monatsbericht über die Verhandlungen der Gesellschaft für Erdkunde. Berlin, May 1859. This paper is by the learned physical geographer, Professor Karl Ritter. It contains observations by Professor Bopp, on the language of the Siah-Pôsh. In my "Researches into the Physical History of Mankind," I have stated the facts known respecting this remarkable people. Vol. iii. part 2.

† Burnouf, Commentaire, Annotations.

‡ "Die heilige Sage und das gesammte Religions-system der alten Baktrier, Meder und Perser, oder des Zendvolks," von J. G. Rhode. Frankf. 1820.

§ Ritter's Erdkunde von Asien: "Iranische Welt."

abandoned their first habitation, and they were led by their patriarch Djemshid through different regions, first to Çughda, or Sogdiana, and at length to Verene, or Persia.* The oldest portions of the Vendidad are fragments of ancient poems, containing a tradition of this migratory march. The most accurate analysis of ancient historical documents thus coincides with the results of philological researches, in bringing the two great Arian races, if not from a common point, at least from almost contiguous regions; whence the Indian branch extended itself towards the east and south, the Persian, or rather the Bactrian, towards the west. The main proof of this conclusion lies in the history of languages, of which I have not room to trace the particulars, and scarcely to mention the principal results. Suffice it to say, that the Zend, the earliest idiom of the Medes, and Persians, and Bactrians, who, as we learn from Strabo† and Nearchus, all spoke dialects of one tongue, is well known to be intimately related to the Sanskrit and the Prakrit, or the ancient language of Hindústan. So intimate is this relation and so well established, that nobody now doubts the affinity of the nations to whom these languages belonged. At the same time the Zend makes a notable approach towards the German, and other languages of the same stock, spoken in northern Europe.‡

After this brief sketch of the common history of the two branches of the Arian§ family, I must describe them separately.

### 1. *Of the Hindús.*

The natives of India have been admirably described in a

---

* "Die Alt-Persischen Keil-Inschriften von Persepolis. Entzifferung des Alphabets," &c., von Ch. Lassen. Bonn, 1836.

† Strabo, "Geog." lib. xv. p. 724. Ed. Casaub.

‡ "Affinité du Zend avec les dialectes Germaniques," par Eug. Burnouf. "Nouv. Journ. Asiatique," tom. ix. 1832.

§ The want of space in the present work compels me to refer the reader to my "Researches into the Physical History of Mankind," where the subject has been considered.

passage of Dionysius the geographer, of which the following is the translation by Mr. Bryant:—

> "To the east a lovely country wide extends,—
> India, whose borders the wide ocean bounds.
> On this, the sun, new-rising from the main,
> Smiles pleased, and sheds his early orient beams.
> The inhabitants are swart, and in their locks
> Betray the tint of the dark hyacinth.
> Various their functions:—some the rock explore,
> And from the mine extract the latent gold;
> Some labour at the woof with cunning skill
> And manufacture linen; others shape
> And polish ivory with the nicest care;
> Many retire to river's shoal, and plunge
> To seek the beryl flaming in its bed,
> Or glittering diamond.
> The rich soil,
> Wash'd by a thousand rivers from all sides,
> Pours on the natives wealth without control."

No summary description which I could make from a collection of different accounts would bear sufficient force of evidence, especially as the physical history of the Hindús furnishes some facts strongly favourable to the opinion which I maintain respecting the origin of varieties. I shall, therefore, cite some statements of facts from unsuspected witnesses. The following are some descriptive accounts of the people of India from various modern travellers.

Le Gentil says, "Les Indiens en général sont beaux et bienfaits; l'œil noir, vif et spirituel; leur couleur est connue; on y voit de très belles femmes, bienfaites, ayant des traits à l'Européenne." "La caste des Bramines surtout est une très belle caste, un très beau sang; dans cette tribu on voit les plus belles femmes, les plus jolis enfans, et tout ce monde a l'air le plus propre."*

The Abbé Dubois, who resided long as a missionary in the Mysore, says that "the colour of the Hindús is tawny, lighter or darker according to the provinces which they

* Le Gentil, "Voyage aux Indes."

inhabit. That of the castes who are constantly employed in the labours of agriculture, in the southern districts of the peninsula, is nearly as dark as that of the Caffres or Negroes. The Brahmans, and people whose profession admits of their working in the shade, such as painters, and many other artisans, are of a lighter hue. A dark-coloured Brahman and a whitish Parriah are looked upon as odd occurrences, which has given birth to a proverb, common in many parts of India, 'Never trust a black Brahman or a white Parriah.' The tint of the Brahman approaches more nearly to a bright infusion of coffee. Their women, who are still more sedentary, and less exposed to the rays of the sun, are still lighter in complexion than the males. In all castes, without exception, the Hindús have the sole of the foot and the palm of the hand much whiter than the rest of the body."

"In general, the Hindús have the forehead small, the face thinner and more meagre than the Europeans, and they are also very much inferior to them in strength and other physical qualities; they are lean, feeble, and incapable of supporting the labours and fatigues which the other race are habituated to. The Brahmans in particular scarcely ever attempt any laborious effort of the body. This feebleness is no doubt occasioned by the nature of the climate, as well as by the quality of the food to which the greater number of Hindús are restricted. In general they eat nothing but seeds, or such insipid matters. The mass of the people cannot obtain rice for their ordinary fare, but are obliged to sell what they raise."

Mr. Orme has remarked that India has been inhabited from the earliest antiquity by a people who have no resemblance, either in their figure or manner, to any of the nations contiguous to them, and that although conquerors have established themselves at different times in different parts of India, yet the original inhabitants have lost very little of their native character.

The great variety of complexion discovered among the

Hindús has been already pointed out as a fact parallel to what is observed in Egypt and Abyssinia. This subject has been put in the true point of view by Bishop Heber, "The great difference in colour," says the Bishop, "between different natives struck me much. Of the crowd by whom we were surrounded, some were black as Negroes, others nearly copper-coloured, and others little darker than the Tunisines whom I have seen at Liverpool. Mr. Mill, the Principal of Bishop's College, who, with Mr. Cowie, one of the chaplains in the Company's service, had come down to meet me, and who has seen more of India than most men, tells me that he cannot account for this difference, which is general throughout the country, and every where striking. It is not merely the difference of exposure, since this variety of tint is visible in the fishermen, who are naked all alike. Nor does it depend on castes, since very high-caste Brahmans are sometimes black, while Parriahs are comparatively fair. It seems, therefore, to be an accidental difference, like that of light and dark complexions in Europe; though, where so much of the body is exposed to sight, it becomes more striking here than in our own country."

That the colour of the Hindús is, however, not independent of the influence of climate, we learn from the fact that the northern people, especially those of high caste, are fair and handsome. The Rajpúts, in the north-western provinces of India, are described by Major Tod, and all other writers, as of tall stature, stout and well-formed persons, with fine features, hooked noses, arched eyebrows, and fair complexions.* The natives of Kattiwar, in the north of India, as we are assured by a writer who has fully and accurately described them, have frequently light hair and blue eyes.†

But the most striking and conclusive proof that the

* Tod's "Rajast'han," vol. i.

† Lieut. M'Murdo's "Account of Kattiwar;" in the Bombay Transactions, vol. i.

complexion of the Hindús is connected with the nature of the climate which they inhabit, is to be found in the colonies of the Indian race, settled at various times in different parts of the high Himálaya, which forms the northern border of Hindústan.

Many Indian families have emigrated at different times from the plains into high tracts in the Himálaya, where some of them have been settled for centuries. The sources of the sacred rivers, the Ganges and Jumna, are well known to be places of great attraction. In the neighbourhood of Jumnotri and Gangotri, situated at their fountains, the Hindús, as we are assured by Mr. Fraser, are very fair, have often blue eyes, and hair and beards curled, and of an auburn or red colour. It must be observed that the climate is in these tracts extremely cold, so that woollen clothes and blankets are required during the night.*

The natives of the valley of Kashmír are Hindús: they speak a dialect of the Hindí, or the native language of Central India. The climate of Kashmír is cool: the country bears fruits similar to those of Europe. The Kashmirians are as fair as the Southern Europeans.†

* "Travels in the Himálaya," by James Baillie Fraser, Esq.

† Fully admitting the gradual change of external characteristics in the Indian, as we proceed from north to south, I would ascribe this great diversity of colour and feature *in the same locality*, to the mixture of races in India; to the supervention of an Arian immigration upon the original Tartar stock. The evidence of this admixture is also to be seen in the languages of India, which exhibit generally a Tartar structure with Indo-Germanic vocabulary. The Tartar structure I believe to be that of every Indian language: the Indo-Germanic vocabulary is predominant in the northern tongues, less so in those of the south, but is not found in the languages of the wild races; see especially the Gonds, of whose idioms we have grammatical notices.

When two races so different come together, the weakest will, perhaps, generally disappear; but in India the parties were so equally opposed, the numerical weakness of the invaders appears to have been so compensated by energy of character and political union, that a fusion has been the result. The institution of caste probably arose from a wish on the part of the conquerors to prevent this fusion.—Ed.

But the Siah-Pôsh,* or the race of Kafirs who inhabit the high region of Kohistan, and the country on the Hindu-Kúsh, called from them Kafiristan, afford the most striking and curious instance of a branch of the Hindú race settled for many centuries in a cold country, and existing under circumstances extremely different from those which surround the natives of Hindústan. The Siah-Pôsh, as it has been proved by Professor Ritter and the celebrated linguist Bopp, from the vocabularies obtained of their language, speak a dialect of the Sanskrit. They are undoubtedly a branch of the Indian race. They worship Mahadeo, but know nothing of the other Hindú gods, and have customs of their own.†

According to the information obtained by the Honourable Mountstuart Elphinstone and Sir Alexander Burnes, the Siah-Pôsh are people of exquisite beauty, with arched eyebrows and fair complexions. A native of their country seen by Burnes at Kabúl was a remarkably handsome young man, tall, with regular Grecian features, blue eyes, and fair complexion. A few other individuals of the same race who have been seen by Europeans had similar physical characters.

FIG. 56.

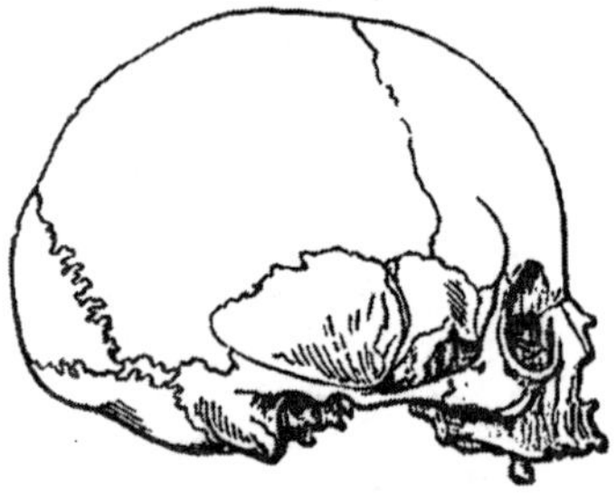

Skull of a Native of Hindustan.

The annexed figure represents the cranium of a Hindú. According to Mr. Martin, from whose work it is taken, with many others in this book, it can only be referred to the type termed Caucasian. "It is of light and delicate

* The language of the Siah-Pôsh is not an isolated one, and Dr. R. G. Latham has suggested the term Paropâmisan for a class of languages more or less allied to it, spoken in the valley of the Gilghit, and as far as the frontier of Little Thibet. He specifies, also, the Lughmani, Der, Tirhai, and more than one dialect of Afghanistan, different from the ordinary Pushtoo language general in the country among the proper Afghans.—ED.

† Ritter and Bopp, *ubi supra*, p. 162.

structure, rather globular, with a prominent occiput, and small cheek-bones."

The adjoined Plate, (6), is the portrait of a Brahman, Ram Ruttun, who was a companion and secretary of Ram-Mohun-Roy. It was taken in Bristol by an excellent artist, Mr. Branwhite, and is a very accurate likeness.

## 2. *Of the Persians.*

A great part of Persia is occupied by half-nomadic people, who wander about the country, living under tents, or cultivate it partially by means of their slaves and dependents: these are the Iliyat, or tribes. Very many of them are not of the Persian race; some are Turkish, others Mongolian hordes or Afghâns, or people of uncertain origin. The towns and their vicinity are occupied by the genuine Persian race, who are every where called not Persians, but Tâjiks. The Tâjiks are, indeed, a people well known, and extensively spread in the East. They inhabit not only the towns of Persia, but of Transoxiana, and all the countries subject to the Uzbek Tartars. Some suppose them to reach to the border of China, at least as far as Tibet.

Sir John Chard the most celebrated of all travellers in Persia, conceived the notion that the old Persian race was an ugly and ill-favoured one, similar to the Mongols, and that the personal beauty for which the modern Persians are not is inherited from Circassian and Georgian concubines. This opinion he probably formed from some of the Iliyat, whom he mistook for Persians. He says, "Le sang des Perses est naturellement grossier. Cela se voit aux Guèbres, qui sont le reste des anciens Perses.* Ils sont laids, mal faits, pesans, ayans la peau rude et le

* It is well known that the Guebres and Parsees are descendants of the Persian fire-worshippers, who, on the conquest of their country by the Moslemin, preferred exile to the abandonment of their ancient superstitions, and took refuge, a part of them in the north-eastern and mountainous provinces of Persia, and a part in India.

teint coloré. Cela se voit aussi dans les provinces les plus proches de l'Inde, où les habitans ne sont guères moins mal faits que les Guèbres, parcequ'ils ne s'allient qu'entre eux : mais dans le reste du royaume le sang Persan est présentement devenu fort beau par le mélange du sang Georgien et Circassien, qui est assurément le peuple du monde où la nature forme les plus belles personnes, et un peuple brave et vaillant, de même que vif, galant, et amoureux. Il n'y a presqu'aucun homme de qualité en Perse qui ne soit né d'une mère Georgienne ou Circassienne, à compter dépuis le roi, qui d'ordinaire est Georgien ou Circassien du côté féminin; et comme il y a plus de cent ans que ce mélange a commencé de se faire, le sexe féminin s'est embelli comme l'autre, et les Persanes sont devenues fort belles et fort bienfaites, quoique ce ne soit pas au point des Georgiennes. Pour les hommes, ils sont communément hauts, droits, vermeils, vigoreux, de bon air, et de belle apparence." "Sans le mélange dont je viens de parler, les gens de qualité en Perse seroient les plus laids hommes du monde; car ils sont originaires de ces pays entre la mer Caspienne et la Chine, qu'on appelle la Tartarie, dont les habitans, qui sont les plus laids hommes d'Asie, sont petits et gros, ont les yeux et le nez à la Chinoise, les visages plâts et larges, et le teint mêlé de jaune et de noir fort désagréable."

Nothing could be further from the truth than the conjecture of this worthy old traveller. It has been contradicted by Sir W. Ouseley, who has shown that all the ancient writers who have occasion to advert to the subject uniformly speak of the Persians and Medes as a remarkably fine and handsome race. They are said to have excelled "κάλλει καὶ μεγέθει"—in beauty and stature; and Ammianus Marcellinus speaks of Persia as a country "ubi fœminarum pulchritudo excellit." A perfect confirmation of this account, which leaves no further evidence to be desired, is afforded by the numerous sculptures on Persian monuments at Istahkar and Hamadan, or Persepolis and

Ecbatana, and other places. The outline of the countenance is here not strictly Grecian, for it is peculiar, but it is noble and dignified; and if the expression is not full of life and genius, it is intellectual, and indicative of reflection. The shape of the head is entirely Indo-European, and has nothing that recalls the Tartar or Mongolian.

The opposite figure affords a good representation of this ancient Medo-Persian physiognomy. It is a specimen engraved by Mr. Morier, displaying the style of countenance portrayed on all the remains of ancient Persian sculpture.

The modern Tájiks, or genuine Persians, called by the Turks Kuzzilbashes, are well known as a remarkably handsome people, with regular features, long oval faces, black, long, and well-marked eyebrows, and large black eyes like those of the gazelle, which among the Oriental people is considered the greatest beauty.

There are several races inhabiting countries near the borders of Persia, and for the most part within the limits of the ancient Iran, who do not belong to the Persian nation properly so termed, but are yet more connected with it than with any other great people of Asia. They must, as I suppose, be referred to the Arian race. They are the Afgháns, the Kúrds, the Balúchi, the Brahúi, the Haikani or Armenians, and, lastly, the Ossetines. I must take a hasty survey of these nations before I proceed to the subjects which are to follow in the order of this work.

### 3. *Of the Afgháns.*

The Afgháns call themselves Púshtaneh, and are termed by the Indians Patans. Afghán is the name to which they are known to Persians, and through them to Europeans. Their speech is the Pushtú,* a dialect derived from the ancient Zend, and therefore a sister language of

* The language is named properly Pakhtu, the second consonant being generally pronounced *kh* by the Afgháns. There is perhaps the same substitution of the guttural for the sibilant as in the vernacular languages

the Persian. It has still some marks of near relationship to the idiom of the Kúrds.

The Afgháns inhabit the mountainous region to the northward of the low country of the Panjáb, or the plain of the Indus. Their proper country is the southern declivity of the great chain of Hindu-Kúsh, the western continuation of the Himálaya and the Paropamisan range: it includes, also, the chain of Soliman, and the table-land to the westward of it. The Afgháns are a rude and warlike people, and are distinguished by their manners and language as well from the Persians as from the natives of India.

The climate of Afghánistan is one of the most delightful in the world. It is dry, as we are informed by Mr. Elphinstone, and the average temperature greater than that of England; the extremes of heat and cold being greater. According to Sir Alexander Burnes, it produces the fruits of England and of Southern Europe—peaches, plums, apricots, pears, cherries, mulberries, vines, and pomegranates; and the groves are stocked with our singing birds, nightingales, blackbirds, thrushes, and doves. The pears and apples of Kabúl are celebrated, and the seasons said to be there delightful. Kabúl itself is more than 6000 feet above the level of the sea. The eastern parts of Afghánistan consist of plains intersected by abrupt chains of hills; the western, chiefly of downs and table-lands, in many parts bleak and cold.

In such a country we might expect to find the people very different from the natives of southern Hindústan. We are informed that the Afghán men are of robust make, being strong and muscular, with high noses and prominent cheek-bones, and long faces. Their hair and beards are mostly black, sometimes brown, but rarely red. Mr. Fraser* describes some Patan, or Afghán, soldiers whom he saw, as having red hair and blue eyes. Mr.

---

of India; the Afgháns mark it by a change in the diacritical points. The Arian character of the language may be doubted.—ED.

* "Journey in the Himálaya," &c. by James Baillie Fraser, Esq.

Elphinstone says, that the eastern Afgháns have generally "dark complexions, approaching to that of the Hindústanees;" while those of the west are of lighter colour, with an appearance of health: but among them, he says, as among the eastern Afgháns, men as dark as the Indians, and others as fair as Europeans, are to be met with in the same neighbourhood,—the fair being the most common in the west, and the dark in the east.* In describing a tribe of Afgháns near Dera, the same writer says,—"The number of children was incredible; they were mostly fair and handsome. The girls have aquiline noses, fine faces, Jewish features. The men were generally dark, though some were quite fair."

The Afgháns are divided into a great number of tribes or clans. The Dúráni are at present the dominant clan, as the Eusofzyi are said to have been in earlier times; the Khyberi and Ghilji are also powerful tribes. I shall not recount the names of the whole number. Though one nation, and little mixed with foreigners, the Afgháns differ very much among themselves in physical character, and the difference is very remarkable. The people who live near the Indus are, as Mr. Elphinstone assures us, black, and resemble the Hindús. The Eusofzyi, who inhabit a high mountainous country in a cool climate, are thus described:—"They are generally stout men, but their form and complexion varies. In those whose appearance is most characteristic of the tribe," says Mr. Elphinstone, "one is struck with their fair complexions, grey eyes, and red beards; by the military affectation of their carriage, and by their haughty and insolent demeanour."

### 4. *Of the Balúchi and Brahúi.*

The Balúchi are a very numerous people of simple pastoral life, who dwell under *ghedans*, or tents, made of black felt and spread over a wicker frame, with which they wander with their flocks over the vast upland of

* "History of Kabúl," by the Hon. Mountstuart Elphinstone.

Kelát, and inhabit most of that great region of eastern Persia which is included between Afghánistan to the north and the Indian Ocean to the south, reaching westward from the Indus to the great Salt Desert. They are a race of Persian Iliyats, and speak a dialect of the Persian language.

Within the same geographical boundaries are the high mountains over which the tribes of the Brahúi are accustomed to wander from one region to another. This race appears to be as nearly akin to the Hindús of the Panjáb as are the Balúchi to the Persians.*

It is interesting to observe that these two races of people, descended one from the Persians and the other from the Hindús, who reversed the local relations of their forefathers, the Balúchi inhabiting in part a warmer and lower country than that of the Brahúi, who are only found in cold mountainous tracts, have also acquired a complexion and physical character corresponding with the climate of their adopted homes. The Balúchi are still a tall handsome race, with good features and expressive countenances; but we are assured that those who dwell in the low plains near the Indus are of very dark colour. The Brahúi, on the contrary, as Pottinger informs us, have short thick bodies, with round faces and flat features, and very many of them have brown hair and beards.†

### 5. *The Kúrds.*

Kúrdistan, or the land of the Kúrds, is the high mountainous tract, intersected by deep valleys, which lies between the great Upland or Plateau of Persia and the

---

* The specimens of their language published by Leech in the Journal of the Bengal Asiatic Society, vol. vii. prove decidedly that it is allied to the languages of the Dekhan, and generally to that class of languages that has been called Tartar. The Brahúi are thus real aboriginal Scyths or Tartars, and their Tartaric features confirm the philological inference.—See Lassen in the Zeitschrift für die Kunde des Morgenlandes, vol. v. p. 358.—Ed.

† "Account of Beloochistan and Sinde," by Lieut. Pottinger.

plains of Mesopotamia. Kúrdistan may be considered as extending from the neighbourhood of the great lakes Urmiyah and Van, southward to the borders of Luristan. Its inhabitants are partly Christians, ancient emigrants from Syria, whence they were exiled on account of their adherence to the Nestorian heresy, who speak the Syriac language; but chiefly semi-barbarous Moslemin, named Kúrds, who are proved by their peculiar idiom to be a branch of the Arian race. They are divided into a great number of tribes, who differ from each other in language, and in degrees of barbarism or improvement. The northern Kúrds are the tribes occupying the four great districts of Bahdinan, Buktan, Hakari, and Rawandiz; the southern tribes are now subject to the Pasha of Suleimaniyah.

The Kúrds are described by the missionary Hoernle, who has given the best account of them, as a strong but coarse people. They are very robust, broad-shouldered, with dark complexion, black hair, small eyes, large mouth, and a wild expression of countenance.

### 6. *Of the Armenians.*

The Armenians are recognised as an Indo-European nation; their idiom is allied to the most ancient dialects of the Arian race, and their early traditions connect them with the history of the Medes and Persians. They are a branch of the same stock with the people of Iran, though separated at an early period, and forming a peculiar people. They adhere with great firmness to the Christian religion and to their ancient church. Of the three millions of souls of which the Armenian race is now supposed to consist, scarcely one hundred thousand have been persuaded to join themselves to the communion of Rome.*

* Versuch einer Geschichte der Armenischen Literatur nach den Werken der Mechitaristen frei bearbeitet," von K. F. Neumann. Leipzig, 1836.

The Armenians are celebrated for the fine form and stature of the men, and for the regularity of features remarkable in both sexes. They have fair skins, with dark hair and eyes.

### 7. *The Ossetines.*

The last branch of the Arian race in Asia are the Ossetines, who dwell on a small part of the Caucasian chain; the greater majority of the inhabitants of this series of mountains being races of people very distinct from the Indo-European.

The Ossetines, as we are assured by Pallas, are a barbarous, predatory race, inhabiting the high and interior country above the Phasis and the Terek. Their language is exclusively spoken by them; but it contains many words and expressions in common with the German, Slavonian, and Persian languages.* "In external appearance they exactly resemble the peasants in the north of Russia; they have in general, like them, either brown or light hair, occasionally, also, red beards. They appear to be very ancient inhabitants of these mountains."

* The Editor has no doubt of the correctness of this attribution: the Indo-Germanic structure of the Ossete grammar appears to him positive evidence, without taking into account the many really Indo-Germanic roots found throughout the language. There is undoubtedly a good deal in it allied to, and, as he is of opinion, borrowed from, the more cultivated Georgian language; and Bopp is much inclined to make even the latter tongue Indo-Germanic too. Dr. R. G. Latham has gone decidedly and far in the other direction, and not only denies a place in the Indo-Germanic class to the Georgian, in which the Editor fully agrees with him, but would also eliminate from this class the Armenian, modern Persian, Kurdish, Baluchi, and those which he terms Paropamisan languages. In regard to Persian, Kurdish, and Baluchi, the Editor has no hesitation in expressing his dissent from this view; his acquaintance with Armenian is too limited to allow him to give an opinion.—Ed.

# CHAPTER VI.

## COLONIES OF THE ARIAN RACE IN EUROPE.

### *European Nations.*

The three celebrated nations whose history we have surveyed appear alone to have possessed in the earliest times the use of letters, and by written monuments to have transmitted to the last ages memorials of their existence. It seems improbable that each of these nations should have become, by a separate process, possessed of this important art; yet those eminent scholars who have laboured with so great success of late in elucidating the Oriental forms of writing, have not succeeded in tracing any connexion between the alphabetic systems of Egypt, of the Phœnicians, the Assyrians, and the Hindús.* From the fact that the art of writing was known at so early a period only in Egypt and in Southern Asia, we must not conclude that other nations were in all respects uncivilised, and destitute of the arts which adorn human life. From the embellishment and perfection of the Greek language, and the history of the Homeric Greeks, we may infer the existence of mental culture, even in ages which preceded the knowledge of letters, or at least the frequent practice of writing.

* This has been doubted, but the Editor is, on the whole, inclined to believe it; only, that instead of Hindús, he would say "people of the extreme east of Asia,"—meaning those who invented the alphabet or alphabets, whatever they were, which gave rise to the Tagala, Korinchi, Lampong, Rejang, and some others still used, or at least known, in the countries east of India. The alphabets now used in India, Burmah, Siam, and Java, are all related, more or less, under manifold disguises and corruptions, to the Devanagari, the oldest forms of which are obviously allied to the Greek, and therefore Phenician. The Egyptian and Assyrian writings were almost certainly, in their origin, symbolical, although the former gradually worked itself down to the Enchorial, and the latter to the Persian,—both alphabetical, but clumsy, and soon superseded. The process, in the Assyrian and Persian at least, was not unlike that by which the symbolic Chinese characters, aided by selection and elimination, produced the alphabets now used in Japan.—Ed.

Having briefly touched upon the history of the three nations who were alone possessed of written memorials, it now remains for me to survey in succession the principal human races who come nearest to them; and here, as I have no particular track which it is incumbent on me to follow, I shall go on, in the first place, to describe that branch of the human family from which the people of Europe are descended. The collective body of the European nations are, as it is now almost universally admitted, a great colony, or a series of colonies, of the Arian or Indo-European race. It would be impossible in the present work, and it would be at the same time foreign to its nature and design, to display in detail the evidence on which this conclusion rests. I shall only collect the results from the heads of argument on which the fact has been established. The proof turns mainly on a comparison of languages. To introduce this subject in a very general point of view to my readers, I shall observe, that, on examining the relations of languages which are said to display marks of resemblance or of connection, two very different series of phenomena are discovered, which lead to very different results. Languages of neighbouring nations, or of nations long and intimately connected by local proximity, by traffic and commercial intercourse, or by political bonds, exhibit marks of such connection in their vocabulary, or in the possession of a great many words in common. Of this description is the extensive resemblance of words between the French and English languages.* In the languages of nations who may have come into a similar nearness of intercourse, while in very different degrees of social culture, when the one people possessed many arts and the knowledge

* I have selected this example as the most familiar instance. It is liable to the objection that the French and English do not belong to originally distinct families of languages. The Anglo-Saxon and the Norman French were, however, so different, that, in a practical point of view, this instance answers my purpose as well as any other.

of very many objects of which the other were wholly destitute, it is evident that a much more extensive resemblance would take place than that which is discovered between the French and English. But this species of resemblance or partial identity in the vocabulary could never approach to what is termed a family relation of languages; that is, such a kind of connection between them as affords proof of a common origin in the nations to which they belong, as in the instance of the English language compared with the German. The first and most important features indicative of family relation between languages is analogy in grammatical structure, and in the laws of combination, or, as we may so term it, the mechanism of speech. Languages supposed to have been originally cognate, have, in some instances, lost almost every other token of relationship except this. It generally happens, however, that grammatical affinity between languages is accompanied by a near resemblance in a certain part of the vocabulary. Occasionally this extends only to a comparatively small number of words; but they are words of a particular class, namely, such as serve to represent the ideas of a people in the most simple state of existence. Such are terms expressive of family relations—father, mother, brother, sister, daughter; names for the most striking objects of the visible universe; terms distinguishing different parts of the body, as head, feet, eyes, hands; nouns of number up to 5, 10, or 20; verbs descriptive of the most common sensations and bodily acts, such as seeing, hearing, eating, drinking, sleeping. As no nation was ever found destitute of similar expressions, and as we know by the observation of facts even more than by the probability of the case, that tribes, however rude, do not exchange their own stock of primitive words for those of a foreign idiom, it may be inferred that dialects which correspond in these parts of their vocabulary were originally one speech, or the language of one people. Now it has been fully demonstrated by those who have devoted themselves to the study of glottology, that both these indications of affinity or family relationship exist between the languages of several races from which the great mass of the population of Europe is derived. This affinity is, moreover, common to the nations of Europe, and to all the Eastern people above referred to the Arian race. Hence the conclusion seems irresistible, that these nations are colonies from Asia, and originally of the Arian stock, which, in an age long preceding our earliest knowledge of European history, spread its branches from the east towards the west and north. Under what circumstances and by what path they originally passed into Europe can only be a matter of conjecture. It is most probable that the northern nations of Europe came into these countries by taking their way through the regions which lie to the northward of the Caspian. From Bactria they must have passed through Turkestan, and advanced between the southern extremity of the Uralian chain and the Euxine towards the mouth of the Danube, whence they spread themselves through Sarmatia towards the north. The nations of Southern Europe, the Italian, and Hellenic, and Illyrian races, may be supposed to have passed into the west by a different route, namely, through Asia Minor, and across the Hellespont or the Bosphorus. It is improbable that any great body of people traversed the intermediate route and crossed over Caucasus, since we know that chain of mountains to have been occupied from remote times by tribes of a race quite distinct from the Indo-European. There is, indeed, as we have seen, among the Caucasian nations, one small tribe of the Arian family, namely, the Ossetines; but they are too insignificant a horde, and too limited in numbers and extent, to have performed any considerable part in the great movements of nations.

If we attempt to enumerate the different nations who are to be considered as ramifications of the Indo-European stock, viewing those as the most ancient which are farthest removed from the centre, or from the path, of migration, we must begin with the Celtic nations in the west of Europe, including the two branches which are represented in modern times, one by the Irish, Scots, and Manks, and the other by the Welsh and Armoricans or Bretons. Next to them in the north of Europe is the Germanic family. It consists, according to the conclusions of the latest and most accurate philologers, of two principal divisions—of the Northmen, ancestors of the Icelanders, Norwegians, and Swedes and Danes; and secondly, of the Proper Teutonic stock in its three subdivisions, which are the Saxon, or Western German, the Suevians or High German, and the Gothic or Eastern clan. The next branch of the Indo-European stock are tribes who speak the dialects of the Old Prussian or Pruthenian language. These dialects are the Lettish, Lithuanian, and the Proper Pruthenian, which of all the languages of Europe bear by far the nearest resemblance to the original Sanskrit. The people who spoke these dialects had a peculiar mythology, and an ancient and very powerful hierarchy, as famous in the north as were those of the Brahmans and the Druids in the east and west. The Slavic, or Sclavonic race, is a fourth Indo-European family: its two great branches are the Western or Proper Slavic, including the Poles, Bohemians, Obotrites, and the tribes near the Baltic; secondly, the Eastern branch, comprehending the Russians, the Servians, and other tribes nearly related to them.* The southern nations of Europe maintain their relation to the same stock. With the exception of the Rasennian or Tuscan people, all the Italian nations belonged to one race; and their dialects,—the Umbrian, the Oscan or Sabine, the Latin, and the Sicelian, or Œnotrian,—are but variations of one speech. The Rasennians were a people of different phy-

* The Slavonic languages are more allied to those called in the text Pruthenian than any other two families of languages here enumerated: they are at least as nearly connected as are the two Celtic families before mentioned, and are by some classed under the common term, Sarmatian.—ED.

sical character from the rest of the Old Italians, and spoke a language which appears to have had little or no affinity to any of the other dialects of the Peninsula. These Italian nations are not, as it has been supposed, descended from a mixture of Greeks, or Pelasgi, with aboriginal barbarians, but form collectively one particular branch of the Arian race; and in respect to the era of their migration from the East, they must be considered as the most ancient of this division. The other races in Southern Europe who belong to the same great stock are the Thracians, the Arnaouts, Albanians, or, more properly, the Skipetars, descended from the Epirots and Illyrians: lastly, the more celebrated Hellenic race.

It would be an interesting question, if there were any data likely to facilitate its discussion, whether the Arian nations found on their arrival in Europe the different countries already occupied by previous inhabitants, or vacant, and affording them a peaceful and undisputed admission. The former hypothesis appears most probable; since we know that the most remote parts whither these nations ultimately arrived were previously inhabited. The Euskaldunes appear already to have possessed Spain before the arrival of Celtic tribes in that country. For if the Celts, as some have supposed, had preceded them, and the Iberian tribes had entered the country at a later period, it cannot be supposed that the latter people, whose military prowess was never comparable to that of the Celts, would have been able to gain possession of the Pyrenean chain, which we know that they inhabited at the era of the Roman conquest.

Spain was the last refuge of this race, who had probably been expelled by the Italian nations and the Celts from Italy and Gaul. In the north of Europe the German nations, or rather the Northmen, found the countries on the Baltic coast already occupied by Jotuns, nations of the Finnish or Ugrian race,—a people, like themselves, of eastern origin, but emigrants of an earlier age, and from a different part of Asia. We shall, in a future chapter, trace their descent from its probable source.

For these and other nations of Europe and of Asia,—distinguished from the Indo-European stock, and not referable to either of the two other primitive races already described,—we stand in need of some distinctive epithet that may assist us in speaking of them in a collective sense, and comparing or contrasting them with nations of the Arian race. I have elsewhere adopted for this purpose the term "Allophylian," which, by its obvious meaning, denotes this distinction, and is preferable to any other name heretofore employed for the same purpose, inasmuch as it can lead us to no erroneous opinion in ethnology.*

Before we proceed to the description of individual nations of either class, it will be useful to compare the Allophylian nations in general with those of the Indo-European family.

The Allophylian nations appear to have been spread in the earliest times through all the most remote regions of the old continent,—to the northward, eastward, and westward of the Indo-European tribes, whom they seem everywhere to have preceded; so that they appear, in comparison with these Indo-European colonies, in the light of aboriginal or native inhabitants, vanquished and often banished into remote and inaccessible tracts by more powerful invading tribes. The latter, namely the Indo-European nations, seem to have been every where superior in mental endowments. Some tribes, indeed, had retained or acquired many characteristics of barbarism and ferocity, but with all these they joined undoubted marks of an earlier intellectual development; particularly a higher culture of language as an instrument of thought as well as of human intercourse. If we inquire into the degree of improvement in the arts of life which the Indo-European nations had attained at the era of the dispersion from their

* Professor Rask used the term Scythian in this sense; but it is uncertain whether many of these nations were Scythians.

primitive abode, or from the common centre of the whole stock, an investigation of their languages will be our principal guide. It gives us strong grounds for a belief that their advancement in useful arts had been comparatively small. The primitive ancestors of the Indo-European nations were probably ignorant of the use of iron and other metals, since the terms by which these are denoted are different in different languages; and must, as it would appear, have been adopted subsequently to the era of separation. Nothing can be more unlike than gold, χρυσὸς and *aurum;* than silver and *argentum;* than *ferrum* and σίδηρος. Other considerations may be advanced to confirm this opinion, that the use of metals was unknown to the earliest colonists of the West. It is plain that the use of letters was entirely unknown to the Arian nations,—to those tribes at least of the race who passed into Europe; and that it was introduced among them in long after ages by the Phœnicians, who claim this most important invention, and certainly have the merit of having communicated it to the nations of the West. But, though unskilled in many of the most useful arts of life, the Arian people appear to have brought with them a much higher mental culture than the Allophylian races possessed before the Arian tribes were spread among them. They had national poetry, and a culture of language and thought altogether surprising when compared with their external condition and habits. They had bards or scalds, *vates,* ἀοιδοὶ, who were supposed, under a divine impulse, to celebrate the history of ancient times and connect them with revelations of the future, and with a refined and metaphysical system of dogmas, which were handed down from age to age, and from one tribe to another, as the primeval creed and possession of the enlightened race. Among them, in the West as well as in the remote East, the doctrine of metempsychosis held a conspicuous place, implying belief in an after state of rewards and punishments, and a moral government of the world. With it was connected the notion that the material universe had undergone, and was destined to undergo, a repetition of catastrophes by fire and water, and after each destruction to be renewed in fresh beauty, when a golden age was again to commence, destined, in a fated time, to corruption and decay. The emanation of all beings from the soul of the universe, and their refusion in it, which were tenets closely connected with this system of dogmas, border on a species of Pantheism, and are liable to all the difficulties attendant upon that doctrine. Among most of the Indo-European nations the conservation of religious dogmas, patriarchal tradition, and national poetry, was confided not to accidental reminiscences and popular recitations, but to a distinct order of persons, who were venerated as mediators between the invisible powers and their fellow-mortals,—as the depositories of sacred lore, and interpreters of the will of the gods expressed of old to the first men, and handed down either orally in divine poems, or preserved in a sacred literature known only to the initiated. In most instances they were an hereditary caste,—Druids, Brahmans, or Magi. Among the Allophylian nations, on the other hand, a rude and sensual superstition prevailed, which ascribed life and mysterious powers to inanimate objects. The religion of fetisses, of charms and spells, and talismans, was in the hands, not of a learned caste, the twice-born sons of Brahma, but of shamans or sorcerers, who, by feigning swoons and convulsions, by horrible cries and yells, by cutting themselves with knives, by whirling and contortions, assumed the appearance of something preternatural and portentous, and impressed the multitude with the belief that they were possessed by demons. Of this latter description were the wizards of the Finns and Lappes, the Angekoks of the Esquimaux; and such are the Shamans of all the countries in Northern Asia, where neither Buddhism nor Islàm has yet penetrated.

[Much of this is hypothetical; but, although materials are accumulating for something more positive, the time is not yet come for pronouncing with any thing like confidence. The author has followed the more usual view;—Dr. R. G. Latham would make the Indo-European or Arian race to be of European origin, and the people of the same race in Asia rather colonists from Europe than *vice versâ*. His main argument is the almost exclusive possession of Europe by tribes universally admitted to be of that race from times beyond the reach even of tradition, and the exceptional condition of their congeners in Asia, where tradition, if not evidence, tends to show that "Allophylian" races once held sway in places since possessed by Indo-Europeans. The editor is of opinion that this argument deserves mature consideration, but he is not prepared to reject the universal belief of the march of man from East to West, the certainty that much of the learning of the West, and all its literature, originated in the East, and the consequent probability that the East is oldest. His own hypothesis, for it is no more, would place the oldest seat of mankind in the East, without wishing to say how far from the borders of Europe; that "Allophylian" tribes were all thrown off in the most ancient times from the great original settled stock, by small detachments or single families, which would lose their social organization and become nomades, by their division into small bodies, and by unlimited space for dispersion. That when the old world was already covered by such scattered tribes, all future emigrations from the central body could only be successfully carried out by large bodies,—large enough to become nations, who would thus retain their original character, and overcome and extirpate the scattered and less organized "Allophylians." By successive waves of such emigrations on a large scale, Europe would be peopled with its present inhabitants; the Allophylian Laps, Fins, and Basques, the Yotuns and Ogres of old fables, giving way before the new comers. It is unnecessary to pursue this further, or we might deduce the unsettled and divided character of the Celts, the earliest because remotest emigrants, from the more extensive contact they must have had with Allophylian tribes in the course of their long journey to the setting sun; the inferior state of the Slavonic nations, the last emigrants, might arise from a similar extended contact in their rear, not being followed up and covered by people in a like social condition with themselves. The decay in the original stock might be attributed to the loss of so many emigrants, by the deterioration arising from an insulated position in the midst of the same Allophylian tribes, who now constitute so large a proportion of the population of the Eastern world. The Editor is quite prepared to find all this treated as a fanciful speculation, and in fact he himself considers it only as one of the many hypotheses which may account for the present state of things.—ED.]

The history of these nations will be the subject of a later chapter. I shall now proceed to say a few words on the physical characters of the Indo-European nations.

### *Physical Characters of the European Nations.*

Incidental notices in the works of Greek and Latin authors give some information as to the physical characters of the ancient inhabitants of Europe. The accounts which we collect in this quarter do not fully agree with the description of the same races in the present times; a considerable alteration must have taken place, if we can trust the testimony of the ancient writers. We must admit the possibility that the physical characters of the nations in question have become changed or modified through the lapse of time, and the influence of external agencies.

There is one other resource from which we may hope to derive information on this subject: I allude to the remains of the dead found entombed in various parts of Europe. A similar research is, as we have seen, one of the most important aids in the investigation of the physical characters of the ancient Egyptians. Sepulchral remains are in Europe much more rare and imperfect than in Egypt; yet there are, if we view them collectively, a great number of such relics, and in some districts they are comparatively frequent. The north of Italy, and especially the country of the ancient Etruscans, abounds in magnificent tombs or places of sepulture. They have been described by Professor K. O. Müller. It appears clearly that these remains, as described by Müller, belong to a people whose physical characters were very different from those of the modern Italians, their descendants. The following observations appeared in a memoir on this subject, contributed by Müller to the "Transactions of the Academy of Sciences at Berlin."*

The countenances of the Etruscans appear to have been of a large and round shape; their eyes large; the nose not long, but thick; the chin strong and somewhat protruding. The figures display in their proportions men of small stature, with great heads, short thick arms, and a clumsy and inactive conformation of body—the "obesos et pingues Etruscos."

The male figures are all beardless, quite smooth and shaven about the chin, dressed in the tunic, or toga, which is sometimes drawn up over the hinder part of the head. On the head they generally wear a wreath of leaves; some hold in their right hand a drinking-cup, and in their left a patera. They repose in an easy posture, a little raised, with their left elbow rested, as if in the attitude of persons who leave the festival of life as well-satisfied guests. The little finger of the left hand is commonly ornamented with a ring. The women lie in the same position as the men: they are clothed with a tunic, some having below their breast a broad girdle, fastened before by wheel-shaped buckles, and with a peplum, which sometimes veils the hinder part of the head. In one hand they hold an apple, or some similar fruit, and in the other a fan. These figures are embossed on the coverings of the sarcophagi, which are formed of stone or of clay. On the clay cover-

* Abhandl. der Berlin. Akad. 1818, 1819.

ings, where a variety of colours is used on the reliefs, these figures are also painted. In them the hair is of a yellow brown colour, and the eyes brown, and the armour and shields of a bluish black, which seems intended to shew that they were made of iron.

Sepulchral tumuli are spread over all the northern and western parts of Europe, and over many extensive regions in northern Asia, as far eastward at least as the river Yenisei. They contain the remains of races either long ago extinct, or of such as have so far changed their abodes and manner of existence that the ancestors can no longer be recognised in their descendants. They abound on the banks of the great rivers Irtish and Yenisei, where the greatest numbers of the then existing people were collected by the facilities afforded to human intercourse. In Northern Asia, these tombs are ascribed to Tschudes, or barbarians, nations foreign and hostile to the Slavic race. The erectors of these sepulchral mounds were equally distinct and separate from the Tartar nations, who preceded the Slaves; for the tombs of the Tartars, and all edifices raised by them, indicate the use of iron tools; and the art of working iron mines has ever been a favourite attribute of the Tartar nations. But silver and golden ornaments of rude workmanship, though in abundant quantity, are found in the Siberian tombs. The art of fabricating ornaments of the precious metals seems to have preceded by many ages the use of iron in the northern regions of Asia.

In the plains where these tombs are found, it is not unfrequent to meet with circles of upright stones, like those which in Europe are termed Druidical, but which are by no means confined to the countries where Druidism is known to have prevailed.

In the western and northern parts of Europe are innumerable sepulchral mounds, or barrows. Many have been examined, both in the British Isles, and in Denmark and Scandinavia. It is much to be regretted that no systematic account has been kept of the results. In this country particularly, nothing has been attempted, in a comprehensive point of view, towards the elucidation of national archæology by similar researches. It appears, however, from late investigations of Professor Eschricht, that the sepulchral remains of ancient European nations may be referred to three periods. The first is the age when tumuli raised over the dead contained no metallic implements or ornaments.* Rings and beads, and other ornaments of amber, in the countries near the Baltic; implements worked from bone, with arrow-heads of flint and fish-bone; celts formed of flint or stone, and other implements manufactured from such materials as we find every where to have been used previously to the invention of metals, are found in various places in tombs of this description. In short, they display a state of rudeness with respect to the knowledge of useful arts not very different from that which prevails in the islands of the Great Pacific.

The osteological characters of skulls and skeletons found in tombs of this era are peculiar: they belong to an earlier race, long ago swept away by one which succeeded it.

It may be worth while to remark that by far the greatest number of barrows opened in different parts of Britain belong to this class. They are so numerous that it is generally believed by antiquaries that the tombs of the Celts prior to the Roman invasion were generally of this kind. There were, however, some belonging to the succeeding class both in Britain and in Ireland.

The second class of tumuli belong evidently to an era subsequent to that which produced the former. In these it is common to find plates of gold, rings of gold or brass, various ornaments of bronze; sometimes swords or blades of brass have been discovered in them, but never tools of iron or indications of sculpture which implied the posses-

* The memoir of Professor Eschricht appeared in the Danish paper entitled "Danske Folkeblad."

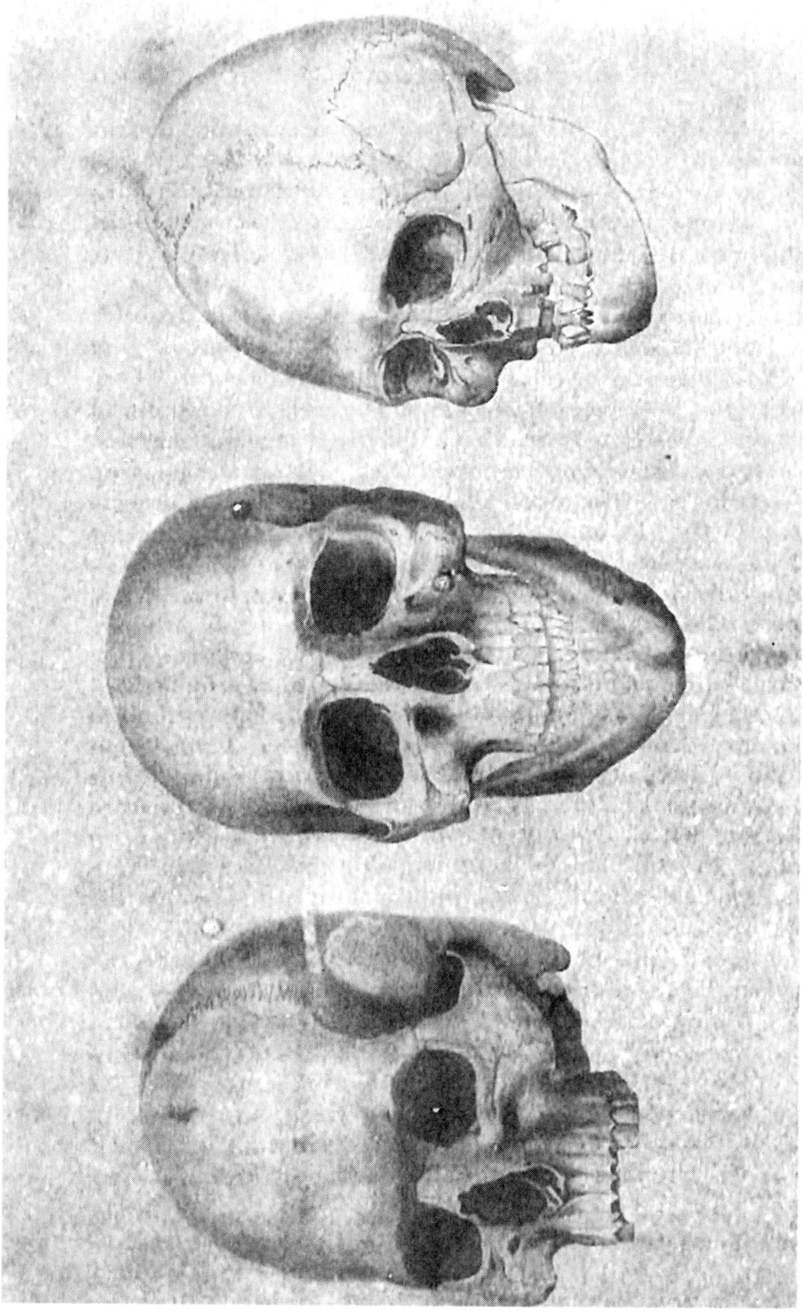

sion of iron tools. A third set of tumuli contain instruments of iron: these evidently belong to a later period than that of the brazen and golden ornaments. The interior arrangements of these different sets of sepulchral mounds are different; but this is a subject beyond our present limits.

The purpose for which I have been induced to offer these observations is to point out the series of osteological remains which may be established by means of them. There seems to be good reason to believe that, by a collection of skulls and skeletons from these different sets of barrows, an historical series may be established, each set displaying the remains of the races of people by whom they were erected. In Denmark, as we learn from the remarks of Professor Eschricht, the barrows of the oldest series contain crania and other bones of a peculiar description. In these the cranium has an ample and well-developed form; the forehead is vaulted, and tolerably spacious; and the nasal bones prominent. At the same time, in a skull of which M. Eschricht has given a description, the zygomata appear large and angular, so that lines drawn from them over the vertex converge, and give the cranium something of the pyramidal form. It may be seen that the eyes were deeply set, with strongly prominent eyebrows, and deep orbits. One of the most remarkable features in these skulls is their round form, approaching to a spherical shape.* In all these points the skulls of this older class make some approach towards the shape peculiar to the Northern Asiatic nations, or to that of the Mongoles and the Esquimaux. The more important circumstances, however, before noted, prevent our referring them to any other type of the human head than the oval and developed shape, which is common to the nations of Europe and of Western Asia. They are probably the crania of Celtic races; in Denmark, those of Cimbrians. The tombs con-

* *Figure I.* on the opposite plate, is the delineation of a cast in the Museum of the College of Surgeons, taken from a skull of this class.

taining ornaments of the precious metals are referred to a later age; but it is uncertain as yet whether they belonged to the same race as the former. Those containing iron implements were subsequent to the immigration of the Germanic nations, who were, as it appears, acquainted at an early period with the use of iron.

There seems to have been no considerable difference between the Celts and Germans in complexion, except that the Germans were more red-haired, while the Celts were flaxen-haired. This has been denied by many writers; but the authority of the ancients is very decided on the subject. I shall not repeat at present the long list of authors whom I have elsewhere cited to this effect,* but shall merely extract one passage from Ammianus Marcellinus, who had lived in Gaul, and therefore must have known of what colour the Celtic people were:—

"The Gauls," says Ammianus, "are almost all tall of stature, very fair and red-haired, and horrible from the fierceness of their eyes, fond of strife, and haughtily insolent. A whole band of strangers would not endure one of them, aided in his brawl by his powerful and blue-eyed wife; especially when, with swollen neck and gnashing teeth, poising her huge white arms, she begins, joining kicks to blows, to put forth her fists, like stones from the twisted strings of a catapult. Most of their voices are terrific and threatening, as well when they are quiet as when they are angry. All ages are thought fit for war, and an old man is led out to be armed with the same vigour of heart as the man in his prime, with limbs, hardened by

* Thus in Virgil's eighth Æneid:—

> "Galli per dumos aderant, arcemque tenebant:
> AUREA CÆSARIES OLLIS, atque aurea vestis;
> Virgatis lucent sagulis; tum lactea colla
> Auro innectuntur: duo quisque Alpina coruscant
> Gæsa manu, scutis protecti corpora longis."

Also Claudian (in Rufinum):—

> "Inde truces flavo comitantur vertice Galli."

cold and continual labour, and a contempt of many even real dangers. None of them are known, like those who in Italy are called in joke Marci, to cut off their thumbs through fear of serving in war. They are as a nation very fond of wine, and invent many drinks resembling it;* and some of the poorer sort wander about with their senses quite blunted by continued intoxication."

The Germans are noted for large broad heads. They were universally celebrated for red hair and blue eyes; and these characters are ascribed to them as uniform, not only by poets, but by writers the most accurate as to matters of fact. Thus Ammianus speaks of the "comas rutilantes ex more" of the Alemanni, the Germans of the Upper Rhine.

It appears certain that the ancient races who peopled the northern and western parts of Europe were all of the xanthous variety of complexion. This is by no means the case with the great mass of people who are supposed to have descended from them. In a poetical chronicle which is supposed by Dr. O'Connor to be the most ancient historical poem existing in the Gaelic language, the bard thus addresses the people:—

"A eolca Albain uile,
A shluagh feta, folt-buidhe,"

rendered:—

"Vos docti Albani omnes,
Vos exercitus peritorum flavo-comatorum."

This is said to have been addressed to the Highlanders at the court of Malcolm III. A.D. 1057. There seems to be a constant tradition that the ancient Gael were a fair-haired race. According to the old legends which contain the story of the Firbolg kings, one of them was named Fiacha Cinnfionnan. Cinfionnan means white heads, and the people, as Keating, the celebrated Irish historian, says, had this designation because most of the Irish of his time were remarkable for their white or fair hair.

* Probably cider, ale, metheglin.

If the Scots of King Malcolm's time were a yellow-haired race, they have forfeited that description, like their countrymen the Caledonians, and like the Germans and Gauls of the Continent. The present Highlanders are by no means generally a xanthous people. In particular districts, and in some valleys in the Highlands, it is noted that most of the inhabitants have red hair; but this is only in limited tracts, where, however, there is nothing indicative of foreign colonisation. The prevalent characters in a great part of the western Highlands are rather dark brown hair, uncurled, with a complexion not very fair, but with grey eyes. A man with coal-black and curled hair, and black eyes, looks singular in a group of the general complexion; and, in places where this variation is frequent, the opposite variety also occurs, namely, a fair skin with red or yellow hair. It seems unquestionable, that the complexion prevalent through the British Islands has greatly varied from that of all the original tribes, who are known to have jointly constituted the population. We have seen that the ancient Celtic tribes were a xanthous race; such, likewise, were the Saxons, Danes, and Normans; the Caledonians also, and the Gael, were fair and yellow-haired. Not so the mixed descendants of all these blue-eyed tribes; the Britons had already deviated from the colour of the Celts in the time of Strabo, who declares that the Britons are taller than the Gauls, and less yellow-haired, and more infirm and relaxed in their bodily fabric. As a proof of this fact, he continues, "We ourselves saw at Rome young men from Britain, who in height exceeded the tallest men there by half a foot, and were crooked in their legs, and not well formed as to the make of their bodies. In their manners," he adds, "they were in some respects similar to the Gauls, in others simple and barbarous."

Nor is this change confined to the Britons and their descendants. The Germans have also varied in their complexion; in the towns of Germany, especially, the people are far from being a red-haired, or even a xanthous race; and, from the fact that this change has been developed chiefly in towns, we may infer that it depends in part on habits, and the way of living, and on food. Towns are much warmer and drier than the country; but even the open country is much warmer and drier than the forests and morasses with which Germany was formerly covered. We must attribute the alteration in physical character to the altered condition under which the present race of people live.*

### *Of the Physical Character of the Slaves.*

No very accurate observations have been made by which it can be determined whether the Slavonians have any peculiar characters distinguishing them from the other European nations; but, if such peculiarities exist, they are of a kind not striking or easily discernible. The various tribes of this race differ among themselves, the variety being apparently in relation to climate and local circumstances; and this variety is much greater than

* The ancient Germans are said to have had universally yellow or red hair and blue eyes; in short, a strongly marked xanthous constitution. This, says Niebuhr, has now in most parts of Germany become uncommon. I can assert from my own observation, that the Germans are now, in many parts of their country, far from a light-haired race. I have seen a considerable number of persons assembled in a large room at Frankfort on the Maine, and observed that, except one or two Englishmen, there was not an individual among them who had not dark hair. The Chevalier Bunsen has assured me, that he has often looked in vain for the auburn or golden locks, and the light cerulean eyes of the old Germans, and never verified the picture given by the ancients of his countrymen till he visited Scandinavia; there he found himself surrounded by the Germans of Tacitus. What can be more evident than that Niebuhr is correct in his opinion that the physical characters of the people have changed? Some alterations in the external conditions under which the race has existed have given rise to a modification in their physical character. The climate of Germany has, in fact, changed since the country was cleared of forests.

any that can be traced between the Slavic nations in general and other Europeans. In the south-eastern parts of their abode, the Slavonians are of dark complexion, with black eyes and hair; this is the fact with respect to the Croats, Servians, and proper Slavonians. The Poles vary in complexion; many of them are of dark eyes and hair, of tall and well-made figures. The northern Russians are very fair; Mr. Tooke observes, that the Russian peasantry have often light-brown, or flaxen, or red hair. Nor is this owing to intermixture with the Finnish race, as some have conjectured; it is too generally spread a character to be ascribed to any such partial and accidental cause. That the xanthous complexion of the northern Russian is not the result of intermixture with foreigners, or particularly with Finns, may be inferred from the fact that other Slavonian nations, who have never lived in the neighbourhood of any Finnish tribe, have, perhaps, in a still more marked degree, the same peculiarity. This may be exemplified in the Slovaks.

The Slovaks are, as we have seen, the old Slavonian inhabitants of Pannonia, or Hungary. They held that country at an early period, and are probably the descendants of the Sarmatæ Jazyges, to whom it belonged in the time of Ammianus. However this may be, they had possession of Pannonia at the period of its invasion by the Magyars, or the Ugrian, or Hungarian people, who gave to it its modern appellation, and who expelled the Slovaks from the central and more fertile plains, into the barren and mountainous tracts bordering on the Carpathian chain, which their descendants still continue to inhabit. The Slovaks form altogether a considerable part of the population of Hungary. A recent English traveller has given us a very minute account of the persons and habits of this race. He says, the Slovaks, in general, are about the middle height, strongly formed, of a light complexion, with broad and coarse features, half shaded by their long flaxen hair. In some particular districts there are found among them singularly fine and handsome men. The peasant women, when young, are sometimes pretty, but hard labour and exposure to the sun soon deprive them of all pretensions to comeliness. In their dispositions the Slovaks are described by the same writer as lazy and indolent; and they are said to be very inferior to the Magyars in energy and activity.

We have a brief account of the persons of the old Antes and Sclaveni from Procopius, which coincides remarkably with this description of the modern Slovaks.

Speaking of the Antæ and Sclaveni, he says: "One language belongs to both nations, which is very barbarous; nor do they differ at all in personal appearance, for they are all of good stature and remarkably robust: as to the complexion of their bodies and their hair, they have it neither very light nor flaxen, nor is it altogether inclined to black; but they are all somewhat red—that is, red-haired."

### *Physical Character of the Greeks.*

It is well known that the remains of Grecian sculpture display the finest and most expanded form of the human skull. It has been supposed, indeed, that the Grecian profile has been exaggerated or drawn from the imagination; but Blumenbach, in a memoir in the "Gottingen Transactions," and in the notes to his sixth decade, has refuted this opinion. He thus describes a Greek skull in his collection:—"Forma calvariæ subglobosa, maxillæ superioris ossibus sub narium aperturis ferè ad perpendiculum coadunatis, jugalibus ossibus modicè et concinnè declivibus, artificum laudatis proxima signis." His Greek skull, and one belonging to the ever barbarous and unintellectual race of Georgians, are said

FIG. 57.

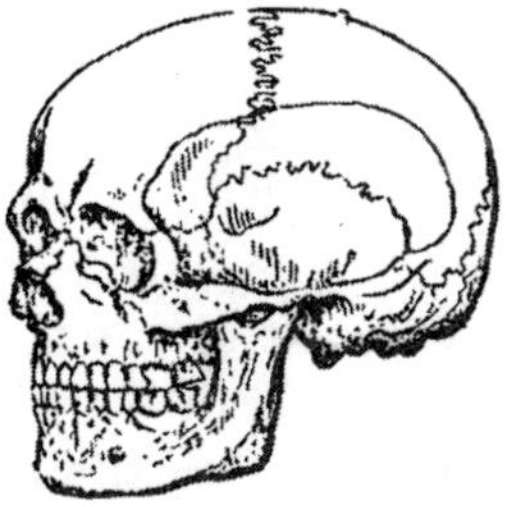

Skull of a Greek.

to be the most beautiful in his whole collection, consisting of 170 crania of different nations.

The preceding outline of a Greek skull will serve to exemplify the form of the head of the modern Greeks.

In the head of the Apollo Belvedere we may probably recognise a good model of the national physiognomy of their ancestors.

FIG. 58.

Head of Apollo Belvedere.

The complexion of the Greeks varied like that of other Europeans, as we know tolerably well from ancient writers. The epithets of ξανθοὶ, πυῤῥοὶ, κυανοχαῖται, γλαυκώπιδες, yellow, red, and black-haired, blue-eyed, and many others, indicate that the same variety of complexion existed formerly among the Greeks which we recognise among other nations in the south of Europe, especially in countries where the climate is varied by differences of situation and of level. It seems that in this respect, as well as in the beauty of form for which the old Greeks were noted, the modern Greeks, their posterity, still resemble them. M. Pouqueville assures us, that the models which inspired Apelles and Phidias are still to be found among the inhabitants of the Morea. "They are generally tall, and finely formed; their eyes are full of fire, and they have a beautiful mouth ornamented with the finest teeth. There are, however, degrees in their beauty, though all may be generally termed handsome. The Spartan woman is fair, of a slender make, but with a noble air. The women of Taÿgetes have the carriage of Pallas when she wielded her formidable ægis in the midst of a battle. The Messenian woman is low of stature, and distinguished for her *embonpoint;* she has regular features, large blue eyes, and long black hair. The Arcadian, in her coarse woollen garments, scarcely suffers the regularity of her form to appear; but her countenance is expressive of innocence and purity of mind. Chaste as daughters, the women of the Morea assume as wives even a character of austerity." The Greek women, in the time of Pouqueville, were extremely ignorant and uneducated. "Music and dancing seemed to have been taught them by nature. The favourable traits of character among the Greeks, in general, are in part attributable to their early education. We are assured that the children are left to grow in full liberty, like the robust plants which adorn their native soil. They are not subjected to the harsh treatment which the children of the lower classes experience in more civilised countries, nor are their countenances expressive of any kind of painful sentiment."

The same writer has described the inhabitants of Sparta. He says, "The Laconians differ in manners and address from their neighbours the Arcadians; the latter carry the scrip and crook, and lead a perfectly pastoral life; the inhabitants of Sparta, on the contrary, fond of combats, are of a lively and restless character, and are easily irritated." M. Pouqueville speaks of the long flaxen hair of the women of Sparta, their majestic air and carriage, their elegant forms, the regularity of their features, animated by large blue eyes bordered with long eye-lashes. "The men," he says, "among whom some are 'blonds,' or fair, have noble countenances; are of tall stature, masculine and regular features." They have preserved something of the Dorians of ancient Sparta.

## CHAPTER VII.

### OF THE FIVE GREAT NOMADIC RACES.

THE great central region of High Asia, whence all the rivers take their source which flow southward into the Indian Ocean, eastward to the sea of Okhotsk and Japan, northward to the Frozen Ocean of Siberia, may be contemplated as a vast upland or plateau, comprehending, perhaps, one-fourth part of the whole area of the Asiatic continent. It is bounded on both sides, namely, towards the north and south by a double series of mountains, each of which, though in rather a low latitude, passes the limits of eternal snow. Of the four longitudinal chains which compose their double barrier, the two southern are the great Himálaya, and to the northward of it, and partly parallel to it, the lofty Tibetan Kuen-lun. In the valley between these two chains, itself the highland of Tibet, of Ladak, and Hlassa, near the sacred lake of Mána-Sarowara, rise the two great rivers of India, Indus and Brahma-putra, which enclose on two sides, and insulate, the whole region of Hindustan. To the northward of Kuen-lun is the great central plain of High Asia, in various parts of which many rivers, which find no exit through such barriers, pour their waters into inland seas. The rivers which flow into Koko-Nor, or the Blue Lake, into Nor Saisan, Lob Nor, and the sea of Balkash, fertilise vast spaces of pasture land; where the primitive nomades of Central Asia fed for centuries their flocks, and multiplied those hordes, which, under their late descendants Attila, Chingis, and Timúr, were destined to change the aspect of human affairs in a great part of the habitable world. To the northward of the central plain, Thianshan, or the Celestial Mountain, and the Golden Mountain, or Altai, separate the Upland from the low region through which the rivers of Siberia take their course to the Icy Sea.

In the centre, or on the borders, of this table-land, which may be termed the Island of High Asia, for such it must have been before the farther subsiding of the ocean laid bare the low plains which lie around its high terrasses on every side, were the abodes, or rather the wandering places, of the five nomadic races: five they may be reckoned, although one of them cannot be traced historically from the plateau. The descent from thence of this first band of nations towards the north-west, can only be inferred by a certain affinity of language and of physical and moral characters between it and the three principal nomadic races. Another member of the group, I mean the south-eastern, belongs not to the central space, but to its Tibetan border. The three members of the central group are the Turkish, the Mongolian, and Tungusian races; the north-western stem is the Ugrian, by some termed the Finnish or Tschudish race; the south-eastern are the Bhotiya, the mountain people, who, on the northern boundary of Hindustan, have appropriated the name of Tartars, though they have no right to that celebrated appellation, which belonged originally to the Mongolian tribe who inhabited the banks of the Lake Bouyir.

To assert that all these nations constitute one race would be to go rather beyond the limits of close induction. But identity of race has often been proclaimed on much inferior evidence. They have occupied—from this remark I must exclude the Ugrian race,*—or rather nomadised

---

* I am inclined to be less cautious than the author, and to believe that in the next note he has made as much allowance as was required, and even more. The larger proportion of the Ugrian race may be still comprised under the observations made, and even the stationary Finn is little more than half reclaimed. The Finn of large towns is really rather a Scandinavian than an Ugrian. It is very doubtful that any of the Arian race are nomadic.—ED.

over contiguous regions from immemorial times; they are nearly on a level in regard to their social state and progress in the arts of life; their moral characteristics, manners, and habitudes, are similar; their religion and superstitions were in early times the same, their physical traits perhaps hardly distinguishable.* Their languages —unless it be that of Bhot, or Thibet, to which this assertion can be applied but doubtfully and in part—though not identical, and long considered by the best informed writers as distinct, yet display, under a careful analysis, such a degree of analogy as proves a distant, but, at the same time, a real, family relationship, and one which may well be comparable to the affinity traced among the most separate members of the Indo-European group.

Though two of their princes, a Turkish and a Mongolian khan, have compiled the history of the Turkish and Mongolian races, and pretend to have traced them from the creation of the world, it cannot be said with truth that the nomadic people of Central Asia possess any memorials of their origin. The compilations of Abulghasi Bahadúr Khan and of Sanang Setzen†—the one a Mussulman, the other a worshipper of Buddha, connect the origin of their respective races,—the first, after the manner of all Mohamedan writers, with the patriarchs of the Old Testament, the other with the incarnate gods, or divine sages of India, celebrated in the fables of Buddhism. It is, however, remarkable that some extensively spread traditions, which are more or less interwoven in all their accounts, and have been collected from very distant times and places, seem to display an obscure reminiscence of the arrival of some fugitive bands from a remote region, who having been saved from destruction by flight into the wilderness, became in process of time the patriarchs of the nomadic

* I limit this remark to the still nomadic races, both of the Turk and Arian stock.

† "Observations sur l'Histoire des Mongoles Orientaux, par Sanang Setzen," par M. Abel Rémusat. Paris, 1832.

races. One of these Sagas relates the fate of a single family, born, or perhaps, if the story were rightly interpreted, suckled, by a wolf, in Turkish Assena, or Tsena,* who became the founder of the Turkish dynasty on Mount Altai. The father of the band, maimed in all his limbs, had escaped from the direful calamities which had overwhelmed his race. Another tradition relates the origin of the Mongols; and this legend was so widely spread, that we find it not only recorded by Rashid-ed-Din and Abulghasi Khan, but alluded to by Sanang Setzen. For ages the Mongolian race had been pent up in the iron-bound valley of Irghænæ-koun. When at length their numbers had so increased that they could not subsist within bounds so confined, they sought and obtained an exit, after melting the iron rocks by fire and with the bellows of seventy forges; and this event was celebrated by a yearly festival till the age of Chingis. The little horde that issued from Irghænæ-koun to conquer the Eastern world were the offspring of two patriarchs, who, many ages before, had there sought refuge. The most numerous and celebrated of all these nations were the Hiong-nu, who possessed an extensive region extending northward from the great wall of China towards the Amúr, and westward from the mountains of In-shan, overhanging the upper course of the Yellow river or Hoang-ho. Their wars against the emperors of the dynasty of Han, coeval with the Christian era, are among the most important events in the history of the great empire. The policy of the Chinese deprecated the enmity of the Tan-shu or sovereign of the Hiong-nu, by giving a princess of the royal house in marriage to their autocrat. The lament of a Chinese lady who had thus become Queen of the Usun, has been deemed worthy of record by the historians

* The Mongolian is Cheno, or Tseno; the Osmanli Turkish has no such word; but, as we find in the old dialects many terms of the Mongol language lost in Osmanli, the word Tsena may probably be old Turkish.—Ed.

of China, and may be cited as characteristic of the manners of these nomadic races :—

> "My kinsfolk have given me away
> Into a foreign land,
> To the chieftain of the Usun.
> He dwells in a miserable hut
> Covered with skins.
> His food is flesh, and milk is his drink.
> When I think of my home,
> Then I long to be a wild goose,
> That I might fly away into my Fatherland."

### 1. *Of the Ugorian or Ugrian race.*

In briefly describing these five nomadic nations, I shall begin with the north-western race, which, as I have said, cannot be actually traced on the central region of High Asia. It is probable that it is one of those races whose expulsion from the high plains by the Hiong-nu is recorded in the history of China ;* but the main proof of its origination from that quarter turns on the analogy of its languages with those of the great central nations. The race who have been termed Ogres, or Ugrian nations, had already left the eastern plateau, and had occupied countries towards the north-west before the earliest accounts. In times long preceding the arrival of the German and Slavic nations in the north of Europe, the Ugrians had possessed all the region extending from the Baltic to the Uralian mountains, and reaching even to the Obi and Irtish, in Siberia. Farthest towards the west were the Finns and Lappes, forming one branch of this race. The people whom the Russians call Tschudes were of the same stock. Farther eastward the name of Ugrians, or Yugorians, predominated. The Ogres were the prototypes of savage monsters, dwellers in forests and mountains, whose name is better known in popular fables than in its historical import. It is, however, the most ancient denomination of this race. The Northmen designated this

* "Researches into the Physical History of Mankind," vol. iii. p. 394, and seqq.

hostile race as monsters and giants; the epithet of Jotnar, or Jotuns, which is of frequent occurrence in the Sagas, had this meaning. Jotuns, among the old poets of the north, as were the Titans of the Greeks, were the enemies of gods and men—creatures of the imagination, symbolical of physical and moral evils. Races of men who were the hereditary and perpetual foes of the Teutonic tribes were also called Jotuns; and this term assumes its historical sense when it is used to designate the barbarous aborigines of Northern Europe, whose conquest or extirpation by a race of happier destinies is celebrated in the early poems of the Scalds. Traces of these older inhabitants of Scandinavia are found in the stories of their warfare handed down from the early historical age. Adam of Bremen, who, during the eleventh century, in the character of missionary, as well as in military service, lived twelve years with the Danish king, Swen Ulfson, has preserved a relation of this kind. "Narravit mihi," he says, "rex Danorum sæpè recolendus, gentem quandam ex montanis in plana descendere solitam et incertum esse unde veniat." "Subitò accedunt; omnem depopulantur regionem." Enemies of civilisation, these barbarous natives of mountains and forests, who were clothed with the skins of wild beasts, and uttered sounds more like the cries of wild animals than the speech of men—"qui ferarum pellibus utuntur pro vestibus, et loquentes ad invicem frendere magis quam verba proferre dicuntur"—dwelt in caves and the clefts of rocks, and issued thence as nightly marauders to perpetrate deeds of blood. By the Icelanders they were termed Jotnen and Thursen—giants and enchanters. That these designations do not belong to the mere creatures of the fancy, such as superstition in later times associates with them, appears from the fact that the historical Sagas deduce the genealogy of many families from a Jotnian ancestry. The early poets, according to Geijer, describe real wars in the accounts of contests against barbarians of the rocks and mountains. In the song of Thiodulf to the honour of Thor, that god is termed the destroyer of mountain-wolves, the overturner of the altars of the Fornjotish idols, the conqueror of Jotuns and Finns. Here an historical name comes forward in connexion with the old term of Jotun to explain its meaning in still earlier use. So Snorro Sturleson, in the "Heimskringla,"' uses Finns and Jotuns as synonymous. The people thus termed are plainly the Skrithfinni, who were described by Procopius as inhabiting Thule in the sixth centuary, and by Paul Warnefrid's son in the eighth, under nearly the same name; and of whom Adam of Bremen reports that they exceeded wild beasts in the swiftness of their flight. They dwelt, according to Adam, towards the north, between Sweden and Norway, especially in Helsingland. He also mentions them in the Wermlands. In the eleventh century, they wandered in the southern frontiers of Norway; in early periods they were certainly in the south of Sweden, where, in a part of Smaland, are still found the local names of Finweden, the field of Finns, Finnheide, and Finnia.

The Finns were, in the time of Tacitus, as savage as the Lappes; but the former, during the succeeding ages, became so far civilised as to exchange a nomadic life for one of agricultural pursuits; while the Lappes have ever continued to be barbarous nomades, as well as the Siberian tribes of the same race, namely, the Woguls and Ostiaks, The Finns, as well as their brethren the Beormahs, or Finns of the White Sea, had probably undergone this change long before the time when they were visited by Otther, the guest of Alfred. When the Finns were conquered by the Swedes they had long been a settled people, but one of curious, and singular, and isolated character.

The eastern branches of this race are the Woguls of the Uralian Mountains, and the Ostiaks on the Obi: from them are descended the Magyars, or Hungarians, a warlike and energetic people, unlike their kindred in the North, in whom a long abode in the centre of Europe has developed the physical and moral qualities of the Arian

Prichard, Nat. Hist. Man.

MALE & FEMALE FINLANDER

London H. Baillière Publisher & at 290 Broadway New York.

race, and whom it has proved to be susceptible of the highest culture. Intermediate between the Uralian nations of the western Finns are various tribes of the same race—Mordwins,* Zyrians, Cheremisses, and Wotiaks, who are termed by Müller, the learned historian of this family of nations, Bulgarian Finns, or Ougres: they were long subject to the Turkish khanate of Bolgari on the Wolga.

### 2. *The Turkish Race.*

The Turkish tribes have been often erroneously termed Tartars. The real Tartars, or rather Tatars, were a people nearly allied, not to the Turks, but to the Mongols, who had their ancient seat in the neighbourhood of the Lake of Bouyir, in the east of Mongolia. All the most learned writers on Asiatic history from the time of De Guignes,

* The Mordwins are principally found in the governments of Simbirsk, Penza, and Saratov, and their number in 1838 is stated to have been 388,111. The Zyrians are in the government of Arkhangelsk, and in 1826 there were in Vologda above 52,000 of this race. The Cheremisses are chiefly in Viatka and Kasan: they numbered 118,952 in 1838. The Wotiaks, chiefly in Viatka, were 161,759. The Woguls and Permians and Ostiaks are still less numerous. The Chuwasses, chiefly in Simbirsk and Orenburg, numbered 145,166; they are in manners and habits like the rest of the Ugrian tribes, but their language is a much-corrupted Turkish. See Koeppen. Mem. St. Petersb. Vol. vi.

Grammars and vocabularies of these tribes have been published, and versions of portions of Scripture have been made of some of them, but all by foreigners. The Zyrianian is distinguished from the others in having once had an alphabet, which was invented for their use by St. Stephen of Perm, who died in 1378. An inscription of ten long lines in the church of Voshemsk in Vologda, is believed to be in the Zyrianian language, but it has not been deciphered, though several of the letters are obvious enough. We have no distinct remains of the Bulgarian language. The whole nation would seem to have migrated to the Danube in the seventh century, to have been gradually absorbed into the Slavonic and Wallachian populations, and to have formed with them an independent State, which continued to subsist during the 13th and 14th centuries. The name still remains borne by the inhabitants of the provinces of Bulgaria in Turkey, whose decided Sclavonic language has received modifications which give it a character distinct from that of cognate idioms: according to Schafarik, a competent judge, it still retains significant traces of a Finnish admixture. See Neumann, Die Völker des südlichen Russlands, Leipzig, 1847, p. 96, 125.—ED.

including M. Abel Rémusat, and Klaproth, and Professor Ritter, are agreed in the opinion, which seems indeed unquestionable, that the Turkish races, now spread through different regions, from the wall of China to the Danube and Adriatic, are of the stock of the Hiong-nu, that powerful and celebrated people who threatened China before the Christian era, and formerly occupied an extensive region, including nearly all the countries now called Mongolia, from the north of China to Mount Altai. After the fall of the empire of the Hiong-nu, they are known in Chinese history by the name of Thu-k'iú, or Turks, and Whey-ou-eul, by Europeans written Huy-hurs, and more correctly Ouigours.* The Ouigours, or Eastern Turks, whose history has been elucidated by Abel Rémusat, are the link of connexion between these more remote nations and the Seljúki and Osmanli Turks who are known to European historians. To trace the affiliation of these tribes in the earliest accounts that remain of them, would occupy too wide a space.† In the present chapter I shall only make a few remarks on their physical characters.

The Turkish nations now existing display two very different types of countenance and of bodily organisation. The nomadic tribes, those who inhabit the ancient abodes of the race, and preserve their pastoral erratic life, have still the physiognomy and general characteristics which appear to have belonged to the primitive Turkish people.

A good specimen of the nomadic Turkish races is furnished by the widely spread race of Kirghis, who inhabit the frontiers of Russian and Chinese empires, and nomadise over vast mountain-plains, from the Lakes Aksakal and

---

* The resemblance of this name to Ogre and Ugrian is obvious. There can hardly be a doubt that the same root is found also in Bulgar and Vogul, and in Vengri and Hungary. The Bolgari probably gave their name to the river Volga.

It may not be too fanciful to connect the English word *ugly* with this root; I am not aware of any other etymon, and the whole race was certainly stigmatised as such throughout the middle ages.—ED.

† See "Researches into the Physical History of Mankind," vol. iv.

Tenghiz or Balkash to the high region of Pamer. I shall cite the description given of the Kirghis by a late traveller, who cannot be suspected of bias to any hypothesis.

Lieutenant Wood, in his account of his journey to the source of the Oxus, has described the Kirghis. He says: "In stature the Kirghis are under the middle size: of a kyl numbering seven men, the tallest was five feet, five and a half inches in height. Their countenance is disagreeable; the upper part of the nose sinks into the face, leaving the space between their deeply-seated and elongated eyes without the usual dividing ridge; the brow immediately above the eye is protuberant, but starts back more abruptly than in Europeans; their cheeks, large and bloated, look as if pieces of flesh had been daubed upon them; a slender beard covers their chin, and in those individuals who have more luxuriant hair the beard has a natural curl. Their persons are not muscular. Their complexions are darkened by exposure to all weathers, rather than by the sun. The women are rather good-looking, and of delicate form, like the Hazaras, and make good wives." He remarks in several places on the ruddy and healthy complexion of the Kirghis females. He says: "The Kirghis resemble the Uzbeks, but the difference between a temperate and a rigorous climate is observable in the well-proportioned frame of the Uzbek and in the stunted growth of the Kirghis. They profess to be related to the Uzbeks, and speak the same language."

Other travellers confirm this account. The missionaries, MM. Zwick and Schill, assure us that the physiognomy of the Kirghis bears a strong and decided resemblance to that of the Mongols. Blumenbach, who has described two Kirghisian skulls in his collection, found the Mongolian characters fully exemplified by them. He has given drawings of two crania (figs. 59 and 60), one of a Kirghis and the other of a Cossack* of the Don, which exemplify these observations. Both have the Mongolian form very fully displayed.

* The word Cossack has rather a social or political than an ethnological value. The Cossacks are generally of Slavonic race, though Turks and Mongols are mixed up with some of their tribes.—Ed.

It would occupy many pages to collect from different travellers the descriptions of all the other nomadic Turkish races. Many accounts of them will be found in my "Researches." The result is, that all the Turkish races who still follow their ancient nomadic life, and wander in the cold and dry deserts of Turkestan, have the so-termed Mongolian physiognomy. Even the Nogays of the Crimea present still much of this character. It is also displayed, to advert to the most distant extremity of the country over which the Turks are spread, in the Yakuti in Eastern Siberia, who live on the lower course of the Lena.

FIG. 59.

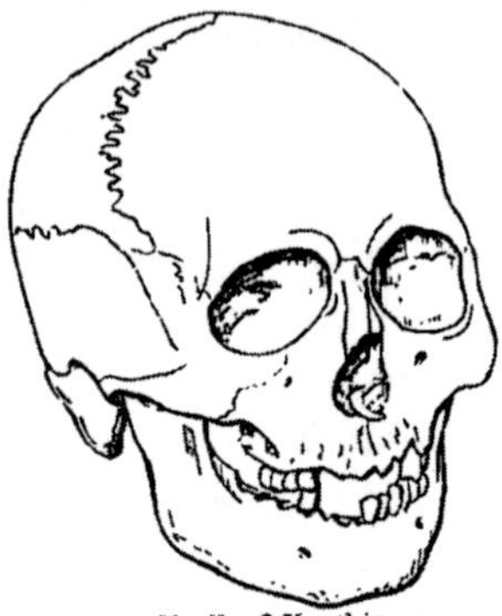

Skull of Kirghis.

FIG. 60.

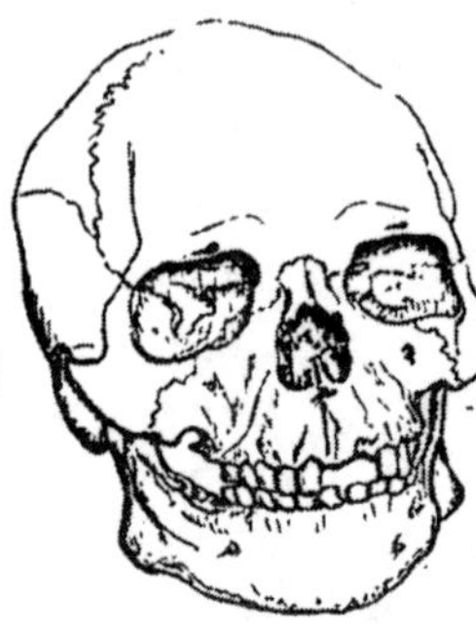

Skull of Don Cossack.

Many writers, not unaware of these facts, and still determined to refer the Turks to a Caucasian stock, attempt to explain their assimilation to the Mongols by supposed intermixture of races. The evidence of language contradicts this attempt. Most of the nations alluded to speak a pure Turkish language, with little or no admixture of the Mongolian. We know besides, from abundant historical proofs, that the Mongols were always a people so small and insignificant in numbers, in comparison of the Turks, as to render this supposition, on an extensive scale, quite inadmissible. The Turkish race was, in fact, aboriginal, in the modified sense in which I venture to use this expression, in the remote regions of Central Asia. They were a people originally akin to the Mongols and Tunguses, and partaking of their physical character.

The Turkish conquests in the West began in the reign of Yezdejird, the last fire-worshipping king of Persia, whom they had assailed before Omar brought Islam to supplant the religion of Ormuzd. We may date the settlement of the Turkish tribes in Mawera'lnahar and Khorasan, and their approximation to the habits of civilised and agricultural nations, nearly from the time of the Hegira.

The Osmanli Turks are in great part descended from the hordes who formed the armies of the Seljukian conquerors of Khorasan. They are the most anciently civilised of the race. The type of their features, and their whole organisation, is in some wholly, and in others nearly, on the European model. This is apparent in the sketch of a Turkish skull, which I here give from Mr. Martin's work. It is, as he says, remarkable for its globular form; it has the forehead broad, and the glabella prominent. The general proportion of the face is symmetrical, and the facial angle nearly vertical.

FIG. 61.

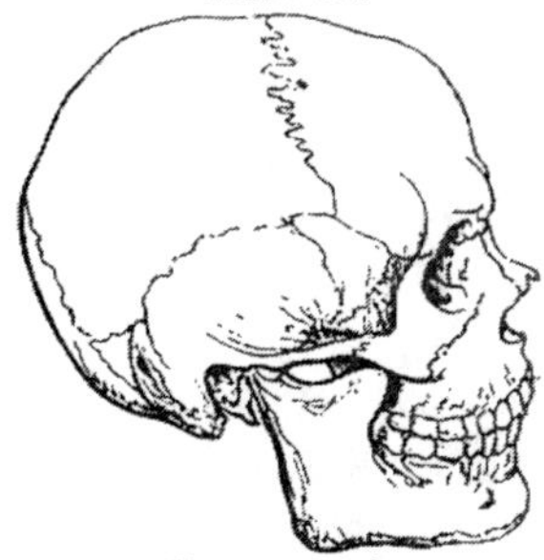

Skull of Turk.

FIG. 62.

Modern Ottoman.

The portrait of an Osmanli Turk on the preceding page affords confirmation of the same remark.

The Tartars of Kasan and some adjoining provinces of the Russian empire are among the most anciently civilised of the whole Turkish race. Their crania, as Blumenbach has abundantly shown, have nearly the European character. The adjoining sketch exemplifies this remark; it is contrasted by Blumenbach with the form of the Don Cossack and the Kirghis.

Fig. 63.

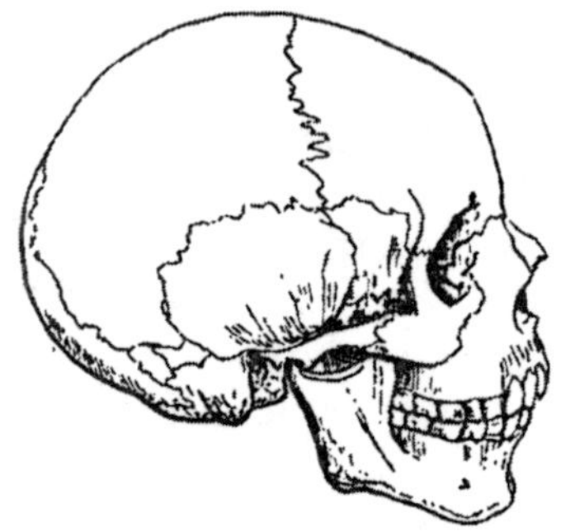

Skull of Tartar.

### 3. *The Mongolian Race.*

The Mongolian race, properly so termed, is generally considered as most strongly exemplifying the broad-faced or pyramidal form of the skull. This character is, however, in reality more fully displayed in the heads of the Esquimaux, and some other nations who wander along the shores of the Icy Sea. But the Mongolian race decidedly belongs to a variety of the human species which is distinguished from Europeans by the shape of the skull. One peculiarity ascribed to the Mongolian race is the globular form of the skull, most remote from that of the elongated prognathous head of the African Negro. But this character is found, as we have observed, in many European races; it has been particularly observed in the crania found in tumuli which are supposed to have belonged to people of Celtic race in the Northern parts of Europe.

The physical character of the Mongols is well described by Pallas. The following observations refer to the Kalmuks, who are well known to be a tribe of the Mongolian nation settled in the plains near the Caspian. "It is easy," says Pallas, "to distinguish by the traits of physiognomy the principal Asiatic nations, who rarely contract marriage except among their own people. There is none in which this distinction is so characterised as among the Mongols. If the colour is set aside, the Mongol has as little resemblance to other people as a Negro has to an European. This peculiar conformation is distinguished particularly in the shape of the skull of the Kalmuks; but the Mongols and the Buriäts have so great a resemblance to them, both in their physiognomy and in their manners and moral economy, that whatever is related of one of these nations will apply as well to the others. The Kalmuks are generally of a moderate height. We find them rather small than large. They are well made; and I do not remember to have seen a deformed person. They entirely abandon their children to nature; hence they are all healthy, and have their bodies well proportioned. They are generally slender and delicate in their limbs and figure; I never saw a single man among them who was very fat.

"The characteristic traits in all the countenances of the Kalmuks are, eyes of which the great angle, placed obliquely and downwards towards the nose, is but little open, and fleshy; eyebrows black, scanty, and forming a low arch; a particular conformation of the nose, which is generally short, and flattened towards the forehead; the bones of the cheek bony ; the head and face very round. They have also the transparent cornea of the eye very brown; lips thick and fleshy; the chin short; the teeth very white: they preserve them fine and sound until old age. They have all enormous ears, rather detached from the head. All these characteristics are observed, more or less, in every individual, and often united in the same person." The following remark, however, seems scarcely to agree with some of these assertions:—"According to the relations of many travellers, one would be led to believe that *all* the Kalmuks have hideous and deformed figures. We see, on the contrary, among the men as well as the women, many round and very pretty faces: we have seen women with such fine and regular features that they would find admirers in all the cities in Europe."

The sketch of a Mongolian skull here inserted was copied from Blumenbach's plates. The oblique position of the eyes in the heads of the Mongolian and other similar tribes is not seen in the shape or position of the orbits themselves: it is produced by the tension of the skin over the projecting cheek-bones, and by the flatness of the space between the eyes. The portrait of Feodor Ivanovitsch, a Kalmuk, who was a painter of some celebrity at Rome, exemplifies this peculiarity and the general expression of the Mongolian countenance.

Fig. 64.

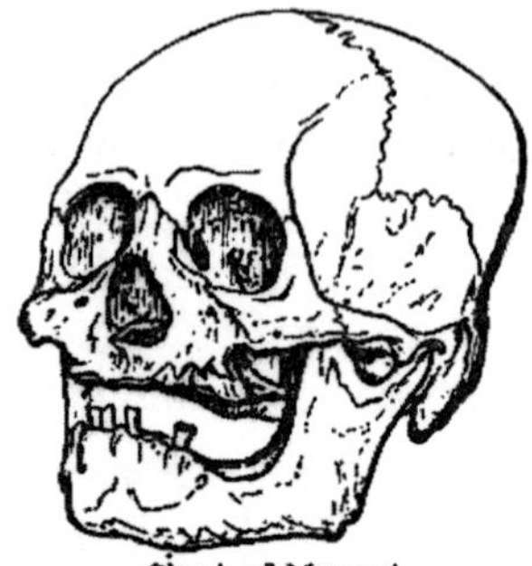

Skull of Mongol.

Fig. 65.

Feodor Ivanovitsch.—A Kalmuk.

### 4. *The Tungusian Race.*

The Tungusians wander over the immense mountainous regions which extend from the Lake Baikal to the Sea of Okhotsk. To the northward, they are dispersed through various countries on the Lena, the Indigirska, Kolyma, and Tungooska rivers towards the Icy Sea; but their proper and original country is probably Daouria, to the northward of Korea and China, where they occupy the districts watered by the Amoor and Usuri rivers. To the northward of the river Uda, they are found on the shores of the great Eastern Ocean. All the tribes of Tungusians within the limits of the Chinese dominion bear the general name of Manchú: they are improperly termed Manchú Tartars. The Tungusians in the dominions of Russia are divided into Dog-Tungusians, Horse-Tungusians, and Reindeer-Tungusians, according to their different habits.

The Tungusians have been a distinct race from very early times. Long before the era of the Manchú empire, which was established in the seventeenth century, nations of the same race appear to have been powerful on the northern frontier of China. From Klaproth's investigations it appears extremely probable that the ancestors of the same Manchús were the people who erected the powerful empire of Kin at the beginning of the twelfth century; and that the Kitans, who established the empire of Liao at the commencement of the tenth, were another nation of the Tungusian race.

The language of the Tungusians is peculiar to themselves. An observation of Klaproth, which he has confirmed by proofs, is here deserving of our attention. He says that the Tungusian, Mongolian, and Turkish dialects display a singular and remarkable connexion between themselves; but what appears yet more striking is the great number of correspondences which the Manchú vocabulary in particular displays with other Asiatic, and still more with European languages.

Pallas, who travelled through Daouria, has described

the Tungusians. He thinks their countenance still more flat and broad than the Mongolian, and more similar to that of the Samoiedes, who belong to the group of Northern Ichthyophagi, presently to be mentioned. I shall cite his description :—

"Leur visage est plus aplati et plus grand que celui des Mongols ; c'est une ressemblance que je leur trouve avec les Samoyèdes. Ils ont peu de barbe ; plusieurs n'en ont point du tout, sans se l'être arrachée. Lors de mon voyage en Daourie, j'avois emmené avec moi un vieillard Toungouse et son fils. Quoiqu'agé de soixante-dix ans, il étoit fort gai, et avoit la peau du visage aussi douce qu'un adolescent. Leur chevelure est noir et longue ; ils la laissent pendre naturellement autour de la tête, à une longueur uniforme. Ils conservent une loupe de cheveux plus longue sur le sommet de la tête, et en forment une tresse pour y attacher leur arc, et le tenir à sec, lorsqu'ils sont obligés dans leurs voyages ou à la chasse de traverser une rivière profonde à la nage."

The Manchú Tungusians, who have been settled in China nearly two centuries, still retain much of the physical character of the nomadic Tungusians ; but this character appears to be in general much softened. Many individuals are there seen belonging to this same race who have an entirely different type of physiognomy. Sir John Barrow, in his description of the Manchús in China, makes this observation :—

"We observed several, both men and women, who were extremely fair, and of florid complexion ; some had light blue eyes, straight or aquiline noses, brown hair, immense bushy beards, and had much more the appearance of Greeks than of Tartars."

### 5. *The Bhotiya Race.*

The Bhotîyahs are the nation, often termed Tartars, who inhabit a great part of Tibet and the Himálayan chain, particularly Bhutan, named from them. They are described as having a strongly marked Tartar or Mongolian countenance; but in vigour of body and in stature they are, according to Mr. Turner, very superior to the nations above described. They are Buddhists, and have peculiar habits, among which is their method of marriage: one woman is generally the wife of a whole family of brothers. It would appear as if this custom is less injurious in a physical point of view than the more frequent sort of polygamy.

The language of the Bhotiyahs is peculiar, and makes a great approximation to the Chinese and other monosyllabic idioms. A vast mass of literature is preserved in it, in the monasteries of Tibet.

[The latest observer of the Bhotiyas, Major Alexander Cunningham, of the Bengal Engineers, in his work just published, entitled "Ladák," has some very exact notices on the physical character of these people. Speaking of Prichard's notice, he says, in p. 292,—"Their superiority in bodily strength is perhaps owing partly to the bracing climate of their elevated country, and partly to the former infusion of Hindu blood. I have had practical proof of this superiority amongst the Botis of Láhul, Kanáwar, and Spiti. In 1846, the short Láhuli women carried with the greatest ease, day after day, the roof of my tent, which the taller and finer-looking men of Kullu and Simla refused on account of its weight. Again, in 1847, the Kanáwari and Spiti women carried loads at which the pampered Simla coolis had grumbled. I have repeatedly seen a box, weighing sixty pounds, carried by girls of sixteen and eighteen years of age over the high passes of Kanáwar."

Major Cunningham's observations do not corroborate the statements of the superior stature of the Bhotiyas over that of other Mongolian people. The result of a very great number of actual measurements of men and women, including the tallest and shortest persons to be found, showed an average height among the more unmixed Bhotiyas of 5 ft. 1·8 in. for men, and 4 ft. 9·35 in. for women, among the working classes. The head men of the villages, who "never carry burdens on their backs, are better fed and better clothed, and (when they travel) usually ride from one place to another, instead of toiling up and down the steep and rugged passes of their native mountains," exhibit an average of 5 ft. 4·5 in. The more mixed Bhotiyas of Kanáwar, who intermarry with Hindus, are taller; and the averages regularly increase from Upper Kanáwar, where such marriages are less common, to Lower Kanáwar, where they are frequent,—as the following table will show (p. 293):—

| | MEN. | | | WOMEN. | | | HEAD MEN. | | |
|---|---|---|---|---|---|---|---|---|---|
| | Tallest ft. in. | Shortest ft. in. | Aver. ft. in. | Tallest ft. in. | Shortest ft. in. | Aver. ft. in. | Tallest ft. in. | Shortest ft. in. | Aver. ft. in. |
| Upper Kanáwar | 5 6·0 | 4 11·0 | 5 1·9 | 5 1·5 | 4 4·5 | 4 9·0 | 5 8·0 | 5 4·0 | 5 5·9 |
| Middle Kanáwar | 5 6·0 | 4 9·0 | 5 2·0 | 5 5·0 | 4 9·0 | 4 11·1 | 5 8·0 | 5 0·0 | 5 5·3 |
| Lower Kanáwar | 5 9·2 | 4 11·0 | 5 4·1 | 5 6·0 | 4 10 0 | 5 1·2 | 5 7·0 | 5 7·0 | 5 7·0 |

"The face of the Boti is broad, flat, and square, with high cheek-bones, large mouth, and narrow forehead. The nose is broad and flat, and generally much turned up, with wide nostrils, and with little or no bridge. The eyes are small and narrow, and the upper eyelids usually have a peculiar and angular form that is especially ugly. The eyes are nearly always black, but brown and even blue eyes are seen occasionally. The inner corners are drawn downwards by the tension of the skin over the large cheek-bones; the eyelids are therefore not in one straight line, parallel to the mouth, as is the case with Europeans, but their lines meet in a highly obtuse angle pointing downwards. This gives an appearance of obliquity to the eyes themselves that is very disagreeable. The ears are prominent, very large, and very thick; they have also particularly long lobes, and are altogether about one-half larger than those of Europeans. The mouth is large, with full and somewhat prominent lips. The hair is black, coarse, and thick, and usually straight and crisp. Bushy heads of hair are sometimes seen, but I believe that the frizzly appearance is not due even in part to any natural tendency to curl, but solely to the tangled and thickly agglomerated matting of the hair consequent upon its never having been combed or washed from first to second childhood" (p. 296).

The following excellent description of a native of Lhassa is given by B. H. Hodgson, Esq.* :—

" *Pénjúr, of Lhassa,* 30 *years old.*

| | Feet | Inches | |
|---|---|---|---|
| Total height | 5 feet | 9½ inches | Rectilinear measurement. |
| Length of head | 0 | 9½ | |
| Girth of head | 1 | 10½ | |
| Crown of head to hip | 2 | 5 | |
| Hip to heel | 3 | 4¾ | |
| Breadth of chest only | 1 | 4 | by curve. |
| Shoulder-point to shoulder-point | 1 | 5 | Rectilinear measurement. |
| Arm and hand | 2 | 6⅛ | |
| Girth of chest | 3 | 0 | |
| Girth of arm | 0 | 11 | |
| Girth of forearm | 0 | 9¾ | |
| Girth of thigh | 1 | 6⅓ | |
| Girth of calf | 1 | 1½ | |
| Length of foot | 0 | 10 | |
| Breadth of foot | 0 | 3 | |
| Length of head | 0 | 7¾ | |
| Breadth of head | 0 | 4 | |

" A fine young man, but low in flesh from sickness, and the muscles flaccid. Colour, a clear ruddy brownish, or brunet, rather deep hued, as dark as any of the Cis-Himálayans, and as most high-caste Hindús, no red on cheeks, which are sunk and hollow. Hair moderately coarse, black, copious, straight, shining, worn long and loose, divided from the top of head. Moustache very small, black. No symptoms of beard, nor any hair on chest: sufficient on mons martis, where it is black, and on armpits also. No whiskers. Face moderately large, sub-ovoid, widest between angles of jaws, less between cheek-bones, which are

* Journal of the Asiatic Society of Bengal, vol. xvii. part 2, p. 222.

prominent, but not very. Forehead rather low, and narrowing somewhat upwards; narrowed also transversely, and much less wide than the back of head. Frontal sinus large, and brows heavy. Hair of eye-brows and lashes sufficient: former not arched, but obliquely descendent towards the base of nose. Eyes of good size and shape, but the inner angle decidedly dipt, or inclined downwards, though the outer is not curved up. Iris, a fine deep, clear, chesnut brown. Eyes wide apart, but well and distinctly separated by the basal ridge of nose, not well opened, cavity being filled with flesh. Nose sufficiently long, and well raised, even at base, straight, thick, and fleshy towards the end, with large wide nares, nearly round. Zygomæ large and salient, but moderately so. Angles of the jaws prominent, more so than zygomæ, and face widest below the ears. Mouth moderate, well formed, with well-made, closed lips, hiding the fine, regular, and no way prominent teeth. Upper lip long. Chin rather small, round, well-formed, not retiring. Vertical line of the face very good, not at all bulging at the mouth, nor retiring below, and not much above, but more so there towards the roots of the hair. Jaws large. Ears moderate, well made, and not starting from the head. Head well formed and round, but larger *à parte post* than *à parte ante*, or in the frontal region, which is somewhat contracted crosswise, and somewhat narrowed pyramidally upwards. Body well made, and well proportioned. Head well set on the neck, neither too short nor too thick. Chest wide, deep, well arched. Shoulders falling, fine. Trunk not in excess of proportionate length compared with the extremities, nor they compared with the trunk and whole stature. Arms rather long; within four inches of knees. Legs and arms deficient in muscular development, from sickness. Hands and feet small and well formed, with instep hollow and heel moderate. Toes not spread, nor splay foot. Mongolian cast of features decided, but not extremely so; and expression intelligent and amiable."]

## CHAPTER VIII.

### OF THE ICHTHYOPHAGI OF NORTHERN ASIA, OR BORDERERS ON THE ICY SEA.

BEYOND the central region occupied by the five great nomadic races above described, are various tribes of people spread over the lower countries of Northern Asia, and over the cold plains which are traversed by the Siberian rivers and border the Icy Sea. These tribes wander from place to place with herds of rein-deer, and support themselves partly by pasturage, and in part by fishing or the produce of the chase. They may be considered as belonging to the same great division of mankind as the Tartar nations, whom they resemble in some leading characters, particularly in the form of the skull. But they differ from those nations in other respects, and bear so much resemblance among themselves that they may well be considered as constituting a particular group, or subdivision, of the human family. I shall here distinguish them by the name of Ichthyophagi, or Fishing Tribes, which describes their habits of life.

### 1. *The Namollos.*

The most remote of these nations are the people termed Namollos.* The Namollos inhabit the north-eastern coast of Asia, from the Bay of Kuliuchin to the river Anadyr. They live in villages dispersed at considerable distances from each other, and feed on seals, dead whales cast ashore, and other gifts of the sea. They are a quiet, timid race. In their persons they are below the middle stature, have flat faces with projecting cheek-bones, small eyes, but generally not compressed and oblique, like those of the Mongols or Tartars. The faces of the women and children are so flat that the nose is scarcely visible.

The Namollos understand in conversation the people

* "Voyage autour du Monde," par F. Lutké. Tom. iii. contenant les Travaux de MM. les Naturalistes, redigé par Alex. Postels.

of Kadjak, and speak, in fact, a dialect of the language of the American Esquimaux. They are a tribe of the race who inhabit the range of the Fox, or Aleutian Islands, the long chain which traverses the ocean to the southward of Behring's Straits. It is difficult to determine from any accessible evidence what was the original country of this race; whether they proceeded in the first place from the north-eastern extremity of the Old Continent to America, or came from the latter in an opposite direction. As the Skrællings, or Esquimaux of Greenland, had not reached that country at the time when the Northmen settled their early colonies in it, it may be conjectured that the progress of the race was from the west, since they had not arrived at the more distant point towards Europe till within the age of history.

The Namollos, as it may also be observed, resemble their neighbours the Tcha-uk-thu, commonly called Tchuk-tchi, in many respects; so far, indeed, that they are often confounded with the latter, and supposed to be one people, for both have hitherto been included under the name of Tchuk-tchi.*

Of the Esquimaux, who are akin to the Namollos, and are either descended from them or are the stock from whence they originally sprang, I shall give an account when I proceed to the American races.

## 2. *The Tcha-uk-thu, or Tchuk-tchi, and Koriaks.*

The Tcha-uk-thu, or Tchuk-tchi, and the Koriaks are tribes of one nation, inhabiting the north-eastern extremity of Asia. The former are the most powerful and independent. Saner informs us that the Tchuk-tchi are a tall and stout people, and hold little men in the utmost

* The Namollos are termed by some writers Stationary or Fishing Tchuk-tchi, and were long confounded with the proper Tchuk-tchi, who are a branch of the Koriaks. The most accurate information concerning them is to be found in the narrative of the Russian voyage, by Capt. Lutké.

contempt. Cochrane says that "the persons of the Tchuk-tchi are not peculiarly large, though their dress, which is clean, but of enormous size, gives them almost a gigantic appearance. They have fair, or clear, skins, but ordinary though masculine features. In conduct they are wild and rude. They have no diseases, and live to a great age. Their language bears no affinity to the Asiatic idioms, though it is understood by the Koriaks. The features of the Tchuk-tchi, their manners and customs, pronounce them of American origin, of which the shaving of their heads, painting of their bodies, wearing large ear-rings, their independent and swaggering way of walking, their dress, and superstitious ideas, are also evident proofs; nor is it less than probable that the Esquimaux and other tribes of Arctic Americans may have descended from them, for several words of their languages are alike, and their dress is perfectly similar."

It seems from this account that the nomadic Tchuk-tchi held intercourse with the American nations, and resemble some of them in manners and in their persons. At the fair of the Tchuk-tchi were two individuals from a nation on the American Continent, termed Kargaules. "They bear," says the same author, "more nearly the features of the Tchuk-tchi than those of the hideous-mouthed inhabitants of the islands in Behring's Straits, though of a browner or more dirty colour."*

---

* The Tchuktchi are described by Matiushkin, in Wrangell's Narrative (Lond. 1840, p. 120), and since then more fully by Lieut. Hooper, in "Tents of the Tuski." Their language is more akin to the Kamchatkan than Klaproth makes it; on the other side, it is strongly contrasted with the Yukagir. The Koriak tribes seem to have encroached from the south, and to have displaced tribes akin to the Yukagirs on one side and Eskimos on the other. Dr. R. G. Latham, to whom the Editor is indebted for this note, states that the Yukagir is more like the Eskimo and North-west American tongues, than to the locally intermediate Koriak: the numeral *maluk*, "two," in some of these languages, as far south as Vancouver's Island, appears in the Yukagir as part of the compounds representing six (=2×3) and eight (=2×4).—Ed.

### 3. The *Kamtchatkans.*

The Kamtchatkans or Kamtchadales, are a people long well known to navigators of the Northern Pacific. They were a numerous people, till they became almost exterminated by the small-pox and other diseases, introduced among them by Europeans.

Only the southern part of the peninsula known by their name is inhabited by this race of people, the northern portion belonging to the Koriaks. The Kamtchatkans call themselves Itelman. By Stoller, who described them with accuracy they were imagined to be of Mongol origin, an hypothesis chiefly founded on a physical resemblance, but which is contradicted by an examination of their language. It appears that they constitute a distinct race, which, however, is divided into four tribes, who scarcely understand each other. They are Shamanists, and a people of rude and squalid manners.

The Kamtchadales are described as a people of short stature, swarthy complexion, of black hair, little beard, broad faces, short and flat noses, small and sunk eyes, small eye-brows, protuberant bellies, and small legs. In all these respects it has been thought that they bear a resemblance to the Mongols.

### 4. *The Yukagers, or Yukagiri.*

The Yukagers are another race very little known, living to the westward of the Koriaks. They inhabit the shores of Eastern Siberia, beyond the Lena, between the country of the Yakuts and that of the Tchuk-tchi, and on the rivers Indigirka, Yana, and Kolyma. They resemble the Samoiedes in their manners. We have a brief account of them in Saner's "Narrative of Billing's Voyage," with a copious vocabulary of their language, which appears to be entirely distinct from all the neighbouring idioms, and to have little or no affinity to any other known dialect.

In the year 1739, the Yukagers were very numerous. The tribes of the Omolon were called Tsheltiere; those of

the Alasey, Onioki; and those of the Anadyr and Anini, *Tchuvantsi* and *Kudinsi*. Wars with the Tchuk-tchi and Koriaks have almost extirpated the race. There was once a numerous nation in the Kolyma called Konghini, the ruins of whose villages, with stone hatchets and arrows, are still found.

The descendants of the Yukagiri inhabit the banks of the two rivers Aniny. They were formerly a formidable and warlike people, and it cost the Russians much trouble to subjugate them; they are now all extinct as a pure race. They are said to be the finest race of people in Siberia; the men well-proportioned, with open and manly countenances, the women extremely beautiful: this applies to the mixed race between the Yukagiri and the Russians. Cochrane assures us that the Yukagiri have the Tartar or Asiatic features, meaning, doubtless, the character of countenance termed Mongolian. In another place, he remarks that they are not very unlike the Yakuti.

### 5. *The Samoiedes.*

The Samoiedes are a wandering race who inhabit the great northern promontory of the Siberian coast, and are spread on both sides along the shores of the Icy Sea, where they live chiefly by fishing and the produce of the chase. They are divided into numerous tribes, who may be said to reach almost from the Dwina and the neighbourhood of Archangel, where some hordes of Samoiedes were found by Le Bruyn, to the Lena, in Eastern Siberia. Their name is said to mean "Salmon-eaters." It occurs in the Russian chronicles as early as 1096; and they are mentioned by Jean du Plan de Carpin, commonly called Plano Carpini, in the account of his journey to the court of the Great Khan, in the beginning of the thirteenth century. The Samoiedes were at that time among the subjects of the Mongolian emperor.

The Samoiedes of the Obi, who may probably be considered as a specimen of the whole race, are said by Pallas to differ entirely in language, as well as in their persons and countenances, from their neighbours the Ostiaks. He adds, "Les visages de ces derniers ressemblent à ceux des Russes, et beaucoup plus encore à ceux des Finois; tandis que les Samoièdes ont beaucoup de ressemblance avec les Toungouses. Ils ont le visage plât, rond et large: ce qui rend les jeunes femmes très agréables. Ils ont de larges lèvres retroussées, le nez large et ouvert, peu de barbe, et les cheveux noirs et rudes. La plupart sont plutôt petits que de taille médiocre, mais bien proportionnés, plus trapus, et plus gros que les Ostiaks. Ils sont en revanche plus sauvages, et plus remuans que ce peuple."

The adjoined portrait is that of a Samoiede. The

FIG. 66.

Samoiede.

breadth of countenance, and the fulness of the cheekbones. occupying a considerable part of the face, are well characterised; but, on the whole, this portrait must be a favourable specimen of the race.

The Samoiedes give themselves the name of Khasova: by the Tungusians they are called Jiandal. It was observed by Strahlenberg, that some traces of their language are discoverable in the southern parts of Siberia, in the countries near Tomsk and Krasnoiarsk; and Pallas has clearly proved that the Samoiedes originated from the southern tracts of the country bordering on the Yenisei and the chain of Sayan. Many facts indicate, as he observes, that these regions were formerly much more populous than they now are; and it will no longer be doubted that the Samoiedes had there their ancient seat, when it is known that the Koibals, the Kamaches, the Motors, the Soiots, and the Karakasses, have the same characteristics as the Samoiedes, and speak their language. The Samoiedes themselves declare that they came from some eastern countries.

It is much to be wished that we could have an accurate description of these tribes of the high region, and could compare it with that of the maritime Samoiedes. Pallas says that they resemble the Tungusians in their physical characters. Klaproth found the same people, under the name of Uriangchai, within the border of the Chinese territories, on the chain of Sayan, which is the eastward continuation of the Altai.

From vocabularies collected as specimens of the idioms of these tribes, it seems likely that they will be found to be allied to the dialects of the Ugrian race, and likewise to those of the nations who inhabit the chain of Caucasus.*

### 6. *The Ainos, or Kurilians.*

The insular race inhabiting the chain of the Kurilian Islands, and a part of the Asiatic coast to the southward of the mouth of the great river Amúr and the Island of

* The celebrated Tartar scholar, Von der Gabelentz, has shown to demonstration the affinity between a dialect of the Samoiede and the Ugrian languages. See Zeitschrift der deutschen morgenländischen Gesellschaft, vol. v. pp. 30-40.—Ed.

Jesso, differs in physical characters from the nations on the northern coast. The climate and situation of these islands are also very different from those of the coast of the Samoiedes. Perhaps we are to attribute to the influence of this cause the great physical difference perceptible between the races of men; for the language of the Ainos has, as Klaproth has shewn, so extensive a connexion in its vocabulary with the idiom of the Samoiedes, and with dialects of some tribes of Caucasus, as to render it very probable that there is a near connexion between all these races.*

The best account of the Ainos that we have yet obtained is to be found in the narrative of Von Krusenstern's voyage. Some particulars respecting them were given by La Pérouse and Broughton. The former of these writers says that "the Ainos are rather below the middle stature, being at most five foot two or four inches high. They have a thick, bushy beard; black rough hair, hanging straight down; and, excepting in the beard, they have the appearance of the Kamtschadales, only their countenance is much more regular. The women are ugly enough; their colour, which is dark, their coal-black hair combed over their faces, blue-painted lips, and tattooed hands, allow them no pretensions to beauty." La Pérouse says that "they are a very superior race to the Chinese, Japanese, and Manchús, and their countenances are more regular, and more similar to those of Europeans." "The inhabitants of the Bay of Crillon were particularly beautiful and of regular features." The same writer adds, that "their skin is as dark as that of the Algerines." Broughton says, "they are of a light copper colour;" but Von Krusenstern declares that they are nearly black.

But the most remarkable circumstance in the physical character of the Ainos is, that though the Eastern

* Professor Pfitzmayer's investigations into the Aino language appear to modify to some extent these statements of Klaproth; but the learned Professor's researches are not yet concluded.—ED.

Asiatics are in general very deficient in hair and almost beardless, they are the most hairy race of people in the world. "Their beards," says La Pérouse, "hang upon their breasts, and their arms, neck, and back, are covered with hair. I observe this circumstance," he adds, "as a general characteristic, for it is easy to find individuals equally hairy in Europe." Broughton declares that their bodies are almost universally covered with long black hair, and that he observed the same appearance even in some young children.

# CHAPTER IX.

## CHINESE AND INDO-CHINESE RACES.

THE vast region of Asia, forming the south-eastern corner of that Continent, which reaches on the sea border from the common mouth of the Ganges and the Brahmaputra to the Hoang-ho, or Yellow river of China, and even further northward towards the mouth of the Amúr, or Saghalien, is inhabited by races of people who resemble each other so strongly in moral and physical peculiarities, and in the general character of their languages, as to give rise to a suspicion that they all belong to one stock. With the rivers which descend from the high country of Central Asia, and pour their diverging waters on all sides, after traversing extensive regions of lower elevation, into the remote ocean, these nations also appear to have come down, at various periods, from the south-eastern border of the Great Plateau; in different parts of which tribes are still recognised who resemble them in features and language.

### 1. *The Chinese.*

The Chinese have long been the most numerous and powerful of these nations. Originally, according to their own historians, a small horde of roving barbarians, who wandered about the forest of Shensi, at the foot of the high mountains of the Tibetan border of China, without settled dwellings, clothed in skins, ignorant even of the use of fire, of which no human race has been discovered to be really ignorant, feeding on insects and roots, more destitute even than the Bushmen or the Australian savages, it was only, if we might credit the childish simplicity of their sacred legends, by listening to the sage counsels of their emperors or patriarchs, that they gradually emerged from this state of barbarism, and, by repeated victories, extended their power over the petty states which now constitute the empire. In the time of Confucius, five centuries and a half before Christ, they had not conquered the country to the northward of the Yang-tse-kiang, the river of Nanking.* The empire of China was probably founded by Shi-hoang-ti, who lived 250 years before our era.† It appears that many of the aboriginal nations of China still inhabit mountainous tracts in the interior. Of these we have no information, except that they are accounted by the Chinese barbarians. They are termed Miao, and Miao-tseu. The Chinese, properly so termed, appear, however, to be of one race, which has exceedingly multiplied. They speak at least one language, though in a variety of dialects.‡

### 2. *The Kooraï, or Koreans.*

The Koreans, if we classify these nations by their affinity of language, should rather belong to the depart-

* Duhalde's "Hist. of China." "Réflexions sur les anciennes Observations des Chinois et sur l'état de leur Empire dans les Temps Reculés," par M. de Guignes fils. Lues à l'Institut de France. Malte-Brun, "Ann. des Voyages," tom. viii.

† See Mr. Davis's outline of the History of China, in his excellent work on the Chinese.

‡ Abel Rémusat, "Mélanges Historiques;" see, also, "Mémoires sur l'état politique de la Chine 2300 ans avant notre ère, selon le Chou-King," par M. Kurz, Nouv. Journ. Asiat.; and "Coup-d'œil Historique sur la Chine," par M. le Professeur Neumann of Munich.

ment of Tartar, or perhaps of Siberian races,* than to the Chinese. But they are subjects of the Chinese empire, and resemble the Chinese in their physical characters. The people of Korea are said to have originated principally from a country to the northward of the Chinese province of Pe-che-li; they were long subject to Japan before Korea was conquered by the Chinese.†

The Chinese, and the Koreans, and the Japanese, belong to the same type of the human species as the nations of High Asia; but it seems among them to have become softened and mitigated, and to display frequent deviation from the character which, if we may believe some travellers, is almost uniform among the Mongols. We are assured by Pallas, that at Maimatchin, on the northern boundary of the Chinese empire, many of the Chinese women have a fair complexion, with fine black hair and *good features.* He adds, that the Chinese idea of beauty is such, that those women are preferred who have the Manchú form; that is, a broad face, high cheek-bones, very broad noses, and enormous ears. We may hence infer that these characters are by no means so general among the Chinese as among the Manchús. M. Abel Rémusat, whose information on everything relating to China was singularly accurate, assures us that the women of the middle provinces have fine complexions, with as great variety of colour as those of the middle countries of Europe.‡ The missionary, M. Gutzlaff, says, that at Tientsin he found the inhabitants more like Europeans than any Asiatics whom he had seen: he seems to include the natives of many parts of the Indian Archipelago. "The eyes had less of that depressed curve in the inner angle which is so characteristic of the Chinese countenance;" the females are fair, and are allowed to walk about.

* Klaproth, "Nouv. Journ. Asiatique," 3. Siebold, "Nachrichten über Kooraï."

† Duhalde; Klaproth; Ritter's Erdkunde, 3, p. 386.

‡ Abel Rémusat, "Recherches sur les Langues Tartares."

The portrait annexed affords a good specimen of the Chinese countenance in general. Its greatest characteristics are breadth and flatness in the suborbital region of the face, outward extension of the zygomatic bones, and an angular position of the eyes. The general character of

Fig. 67.

Chinese.

this physiognomy is described accurately by Dr. Siebold, in his account of the people of Korea. He says,—

"L'ensemble de leurs traits porte, en général, le caractère de la race Mongole: la largeur et la rudesse de la figure, la proéminence des pommettes, le développement des machoires, la forme écrasée de la racine nasale, et les ailes élargies du nez, la grandeur de la bouche, l'épaisseur des lèvres, l'apparente obliquité des yeux, la chevelure roide, abondante, d'un noir brunâtre ou tirant sur le roux, l'épaisseur des sourcils, la rareté de la barbe, et enfin un teint couleur de froment, rouge jaunâtre, les font reconnaître, au premier abord, pour des naturels du nord de l'Asie. Ce type se retrouve chez la plupart des Coréens que nous avons vus, et ils conviennent eux mêmes que c'est celui qui distingue le mieux leur nation." The deviations, however, from this form were so considerable in other individuals, as to give the writer a suspicion of the co-existence of two intermixed races. He says that "Le nez écrasé près des canthus internes, et terminé par de larges ailes, les yeux obliques, les canthus internes très éloignés l'un de l'autre, et les pommettes saillantes," are marks of the race first described. "Mais lorsque la racine nasale est élevée, lorsque le dos du nez se prolonge en ligne droite, la figure du Coréen se rapproche déjà du type des peuples d'origine Caucasienne, et la conformation des yeux ressemble davantage à celle des Européens : les pommettes s'effacent alors, et le profil fortement dessiné, qui devient plus apparent, contraste surtout avec celui des Mongols. A' mesure que la physionomie se rapproche de celle de la première des deux races Coréennes, la barbe est plus légère ; elle est plus épaisse chez les individus de la seconde, le sommet de la tête est moins aplati, le front, au lieu d'être renfoncé, offre des lignes droites et pures, et tout leur aspect physique révèle une noblesse qu'on est loin de trouver dans les traits grossiers des Mongols." *

The conjecture of two co-existent races in this region is altogether without support, and it is extremely improbable, because the characters of nations long intermixed are known to amalgamate and become in time blended. There is no reason for doubting that the phenomena described are the result of spontaneous deviation. We have seen that a variety of formation precisely corresponding was observed by Sir J. Barrow among the Manchú natives of China. If, in every such instance, we are to attribute variations in physical character to intermixture of races, we must gratuitously assume the existence of two or three

* "Voyage au Japon," par M. le Docteur Siebold.

distinct races of men in every nation, and in almost every family.*

### 3. *The Japanese.*

The Japanese belong to the same type as the Chinese; they resemble them in many particulars. They owe to China their civilisation, their literature, and, at least, one of their popular religions. The figure from Siebold given below represents the most general form of the Japanese physiognomy.

FIG. 68.

Ko-tsching Dschang.

* Such instances as that mentioned in the text are of a nature to confirm the views regarding mixture of races which the Editor has stated in the preface; they abound wherever, either from the commercial character or wandering habits of a race uncombined with any disinclination to form unions with tribes of different races, or from their border locality as in Corea, such mixture of races would be expected.—ED.

The following passage from Siebold's work on Japan describes the varieties of figure and complexion which he observed in Kiu Sin, one of the great isles which form the empire of Nippon, or Japan.* We may observe that the colour of the hair is often brown or red, though the uniform black has been laid down as a characteristic of the race to which the Japanese are referred.

"La population du Fizen, comme celle de toute l'île de Kiu Sin, se divise en habitans des côtes, de l'intérieur, et des villes, qui diffèrent entre eux par l'aspect physique, la langue, les mœurs, et le caractère. Les côtes et les îles innombrables qui les avoisinent sont habitées par des pêcheurs et des marins, hommes petits, mais vigoureux, d'une couleur plus foncée que celle des autres classes. La chevelure, plus souvent noire que brun-rougeâtre, est crêpue chez quelques individus, qui ont aussi l'angle facial très-prononcé, les lèvres gonflées, le nez petit, légèrement aquilin, et renfoncé à la racine. L'adresse, la persévérance, l'audace, une franchise qui ne va jamais jusqu'à l'effronterie, une bienveillance naturelle, et une complaisance qui touche à la soumission; tels sont les traits caractéristiques de ces habitans des côtes.

"Ceux de l'intérieur de Kiu Sin, qui se vouent en grande partie à l'agriculture, sont d'une race plus grande, reconnaissable à sa figure large et aplatie, par la proéminence des pommettes et la distance des canthus internes, à son nez gros et très écrasé, à sa grande bouche, à ses cheveux d'un brun foncé tirant sur le brun rougeâtre, et à la couleur plus claire de sa peau. Chez les cultivateurs qui journellement s'exposent à l'air et au soleil, la peau devient rouge: les femmes, qui se préservent des influences atmosphériques l'ont ordinairement blanche, et les joues des jeunes filles brillent même d'un vif incarnat."†

* Nippon is the Japanese pronunciation of the Chinese name Yepuen: Japan is the English reading of the same Chinese syllables.—Ed.

† These passages are taken from Dr. Siebold's great work, entitled "Nippon Archiv, zur Beschreibung von Japan und dessen Neben- und

The annexed profile of a Chinese skull appears to differ very little from the European type. In plates III. and IV. the reader will see a correct delineation of the front view, and likewise of the basis, of a Chinese skull. In this the pyramidal shape of the front view is strongly marked, as well as the round outline of the basis, the transverse diameter bearing a considerable proportion to the longitudinal, and the zygomatic bones being large and round. At the same time, by comparing these figures with those of a native American and of a South African, in the same plate, an attentive observer will see a sufficient proof that none of these characters are specific. Though taken from races of men the most widely separated, and two of them, the Chinese and the woolly Africans, supposed to belong to those varieties of mankind which recede most widely from each other, these three crania bear to each other in their general contour an extraordinary resemblance.

FIG. 69.

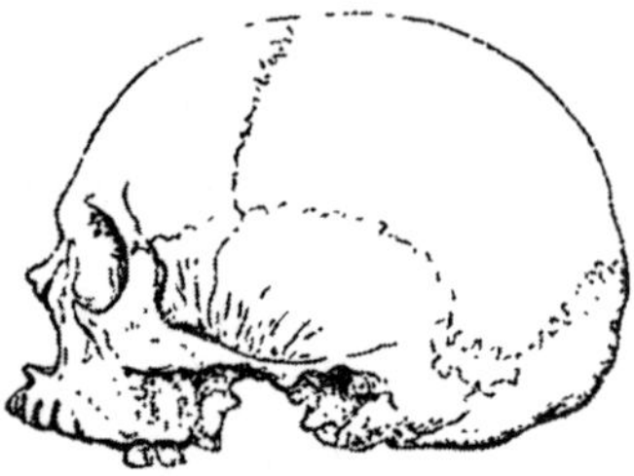

Skull of a Chinese.

4. *Races of the Indo-Chinese Peninsula.*

The great projecting land which constitutes the Peninsula of India beyond the Ganges is formed by several chains of mountains branching off towards the south-east, from the continuation of the Himâlaya towards the Gulf of Tonquin where the Himâlaya terminates. Between these lofty chains are several considerable rivers, which issue from the same high mountainous barrier, and irrigate the long valleys, the ancient abodes of several remarkable nations. The languages and the physical characters of these nations give reason to believe that they all originally issued from the same region as the Chinese. They may be divided,

Schutzländern," &c. French edition. The annexed plate represents a group of Koreans from Siebold's work.

however, into two classes; one consisting of the more ancient inhabitants of the southern districts of the peninsula, who, in comparison with the second class, may be considered as aborigines;* they now inhabit principally mountainous tracts in various parts of the peninsula, and appear to have been expelled from the more level and fertile countries, now occupied by the tribes who belong to the second class. These last are more civilised, and partake more or less of the refinement and peculiar habits

---

* The following notes, extracted from the Report of an Expedition sent to Cochin-China by Warren Hastings in 1778, which was not published until 1852, when it appeared in the pages of the Sincapore Journal, may be of some interest:—"The aborigines of Cochin-China are called Moys, and are the people which inhabit the chain of mountains which separate it from Cambodia. To these strongholds they were driven when the present possessors invaded the country. They are a savage race of people, very black, and resemble in their features the Caffrees.

* * * * * *

"The Cochin-Chinese bear evident marks of their being derived from the same stock as the Chinese. They resemble them in their features, and most of their manners and customs. Their religion is the same; their oral language, though different, seems formed upon the same principles, and they use the same characters in writing. They are a courteous, affable, inoffensive race, rather inclined to indolence. The ladies are by far the most active sex; they usually do all the business, while their lazy lords sit upon their haunches, smoking, chewing beetle, or sipping tea. Contrary to the custom of China, the ladies are not shut up; and, if unmarried, a temporary connection with strangers who arrive in the country is deemed no dishonour. Merchants often employ them as their factors and brokers: 'tis said the firmest reliance may be placed on their fidelity."

We may observe generally that the test of language shows in this race of people a division into two families; not such a distinction as that existing between the speakers of Indo-Germanic and Tartar tongues, but rather such as exists between the more distant classes of Tartar languages. It may be stated that juxtaposition and a community of religion have combined to produce secondary resemblances; giving rise to borrowings from each other, and from a foreign source, the Pali sacred books, which make the Indo-Chinese idioms more alike externally than really; an amount of likeness (but far more pervading) which may be seen between two such unlike languages as French and German; both taking their ecclesiastical nomenclature from Greek, and the German, unnecessarily perhaps, using the terms and expressions of French literature.—ED.]

of the Chinese, and are all subject to the Lamaite priesthood, and follow the worship of Buddha or Fo according to the Chinese form. They are considered in the countries which they have occupied as Chinese colonists, though they differ from the people of China in language, and must be considered as separate nations. It must be observed that all the languages of the races in this Indo-Chinese Peninsula are known to belong to the same group with the Chinese, being of the kind termed monosyllabic. To this last or more civilised class of nations we must refer the race of Anam, in Tonquin and Cochinchina, on the eastern side of the peninsula, the Laos or Lia, who were originally a branch of the same race with the T'hay, or Siamese, tribes of people who occupy all the central and inland parts, and the Barmah, or Rukheng, or Arakan race to the westward, reaching to the Bay of Bengal. To the aboriginal or more ancient class belong the Tchampa to the southward of Anam, the Khomen, or Kambojans, to the southward of the Laos, the Môn, or people of Pegu, to the southward of Barmah, as well as many other races of mountaineers in the interior. In this enumeration I have purposely omitted the nations of the Malayan Peninsula; they are in some respects a distinct class of nations, and will be considered when I proceed to the insular races, with whom they are more connected in history than with the other Indo-Chinese nations.*

The great analogy in the forms of words, and in the fundamental rules of grammatical structure, between these languages and the Chinese and the Bhotiya,† give rise to

* "Researches into the Physical History of Mankind," vol. iv.

† The Burmese language belongs certainly to a class somewhat different from those of the other nations of the Ultragangetic peninsula; it approaches rather to the Bhotiya than to any other idiom known to the Editor. There is, it is true, the connecting link of monosyllabism, and some mutual borrowing of words: the truth appears to be, that the Chinese and Bhotiya are extreme opposing languages of the monosyllabic class, and that Burmese is more allied to the latter language, while the other tongues of the peninsula, the Malay excepted, belong to the Chinese branch.—Ed.

a strong suspicion of original identity. If we were at liberty to hazard a conjecture as to the origin of these nations, it would be that all the people who inhabit the low countries of south-eastern Asia, from the mouth of the Amúr, or at least from that of the Hoang-ho, southward and westward as far as the Brahmaputra, are offsets from one of the great nomadic races of High Asia, namely from the Bhotiya, who occupy the southern margin of the Great Central Upland.

The best description of all these nations is that of an enlightened and philosophical traveller, Mr. Finlayson. The following summary of observations is intended to apply to the various races above mentioned, and in general also to the Chinese, who are regarded by Mr. Finlayson as the prototypes of the whole group.

He begins by observing that the characters of all these tribes are not uniform and constant. He says, "A multitude of forms are to be seen in every nation not referable to any particular family or variety of the human race." This is quite sufficient to disprove the existence of distinct races. He adds, "For our present purpose we must select such only as possess the peculiar form in the most *characteristic degree.* But as the particulars of this form are not always developed in a full degree in all, we must collect from a multitude of instances what appears to be the predominating tendency.

"The stature is nearly similar in all these tribes; the Chinese being, perhaps, a little taller, and the Malays lower, than the others. In all, it is below the European standard; the average height of the Siamese is five feet three inches. The complexion in all this group of nations is lighter than that of most Asiatics on this side of the Ganges, by far the greater number being of a yellow colour, which in the higher ranks, and particularly among women and children, they heighten by cosmetics, so that their bodies are often rendered of a golden colour. The texture of the skin is remarkably smooth, soft, and shining.

"The whole race displays a remarkable tendency to obesity.* The nutritious fluids of the body are directed towards the surface, distending and overloading the cellular tissues with an inordinate quantity of fat. The muscular textures are in general soft, lax, and flabby, rarely exhibiting that strength or development of outline which marks the finer forms of the human body. In labourers and mechanics, particularly among the Chinese, the muscular parts occasionally attain considerable volume, but very rarely that hardness and elasticity developed by exercise in Europeans. A first aspect gives a false estimate of their physical power. In some the limbs often equal those of Europeans, and particularly the thighs; they may be said to form a squat race.

"The face is remarkably broad and flat; the cheek-bones prominent, large, spreading, and gently rounded; the glabellum is flat and unusually large; the eyes are in general small; the aperture of the eye-lids, moderately linear in the Indo-Chinese nations and the Malays, is acutely so in the Chinese, bending upwards at its outer end; the lower jaw is long, and remarkably full under the zygoma, so as to give to the countenance a square appearance; the nose is rather small than flat, the alæ not being distended in any uncommon degree: in a great number of Malays it is largest towards its point; the mouth is large, and the lips thick; the beard is remarkably scanty, consisting only of a few straggling hairs; the forehead, though broad in a lateral direction, is in general narrow, and the hairy scalp comes down very low. The head is peculiar; the antero-posterior diameter being uncommonly short, the general form is rather cylindrical; the occipital foramen is often placed so far back that from the crown to the nape of the neck is nearly a straight line. The top of the head is often very flat. The hair is thick, coarse, and lank; its colour is always black. The limbs are thick, short, and stout, and the arms rather out of

* From this remark the Cochinchinese are afterwards excepted.

proportion to the trunk: the arms, particularly in Malays, are uncommonly long. The foot is in general small; but the hand is much longer than that of the Bengalese. The trunk is rather square, being nearly as broad at the loins as over the pectoral muscles. There is in this respect the greatest difference between them and the inhabitants of India, who are in general remarkable for small waists. The diameter of the pelvis is particularly large, and the dimensions of the cavity would appear to be somewhat greater than in other races."*

From this account of their form, they would appear to be calculated for toilsome and laborious exertions; but they have not the energy of European labourers: the greater number are distinguished for mechanical skill and patience rather than for mental capacity; others are equally remarkable for indolence and aversion to labour.†

* Finlayson's "Embassy to Siam and Hue," p. 230.

† Some valuable papers by Mr. Hodgson, published in the Bengal Asiatic Journal, 1853, show a decided affinity between the languages of Tibet and the Himálaya mountains, and those of the Indo-chinese borderers; and argue with much ability and success for some connection between these and the languages of the Caucasus, as previously shown by Dr. R. G. Latham, whose researches are cited by Mr. Hodgson with high approbation. Mr. Hodgson's first comparisons are founded on vocabularies of the Burmese, Khyeng, (a rude people on the Yomá hills, in Arakan and Burmah, numbering about 27,000), Kumi, and Kami (smaller adjacent tribes whose language has been recently reduced to writing), Mrú (a still smaller tribe nearer to Chittagong), Sák (a very small tribe adjacent), Siamese, Mon or Pegu, Shan, and Tung-Chu, in Tenasserim, (printed in Journal, No. 1, 1853,) collated with Himalayan vocabularies printed in the number for December 1847. In the following paper, he collects the words of several Sifan and Tibet vocabularies, printed in No. 2, with Caucasian words supplied by Rosen; and enters into a more detailed analysis of some grammatical features of the Circassian and Gyarung languages, the latter spoken on the Chinese frontier of Tibet. These collations are strongly in favour of the connections for which Mr Hodgson contends, and further comparisons, continued more largely in a third memoir, carry the alliance to a class of languages where I was not prepared to find an affinity, viz., the Malayan class, more especially the Tagala, the prodigality of whose affixed, infixed, and suffixed particles is rivalled, if not surpassed, by what is found in the Gyarung. I must confess my unwillingness to receive this last connection; I believe

## CHAPTER X.

### OF THE ABORIGINAL RACES OF INDIA.

I PROPOSE to include under this term a variety of distinct races, or rather numerous tribes of people having different languages, and betraying no evident indications of mutual affinity, who are spread through various parts of India. All these races are distinct from the Hindús, who belong to the Indo-European or Arian stock, and they were probably spread through the countries which they now inhabit, though perhaps thinly scattered, long before the ancestors of the Hindus first passed the river Indus. Their languages, as far as they are known, are in construction quite distinct from the Sanskrit and its sister idioms. Several of them, as the Tamul and its cognates, bear in this respect a greater resemblance to the languages of the Tartar nations. The physical characters of these nations are not all according to one type: some resemble the Hindús considerably; others approximate to the Indo-Chinese form. It must be allowed that the constituting of such a department of nations indicates the imperfection of ethnology; but these races are too numerous to be separated into so many classes. Besides, many of them certainly belong to a smaller number of groups, and, until their languages shall be better known and analytically compared, we cannot hope to classify them with any degree of accuracy.

In these groups I shall make subdivisions, and comprehend in each the following nations:—

that glossarial resemblances, and even structural propinquity, may exist *detached* in cases where no alliance can be supposed; as for instance, in the Surgut dialect of Ostiak, which is pervaded by a principle very much like the Indo-Germanic Umlaut, unknown to other Ugrian languages, and quite distinct from the Tartar vocalic harmony; but I hesitate to dissent from so acute an observer as Mr. Hodgson. His other deductions I am prepared to admit, and to find them continued throughout Asia, with the exception of the Semitic area, Persia, and perhaps Armenia; I am equally prepared to find them extended to New Holland.—ED.

1. The Singhalese, including the proper Singhalese, the Kandians, the Vaidas, and in short, all the inhabitants of the Island of Ceylon who do not belong to the Tamulian race.

2. The Tamulian race, inhabiting part of Ceylon and the greater part of the Dekhan, or of the Indian Peninsula. The proper Tamuls are in the southern parts; but nations separated from them, yet speaking dialects of the same language, extend almost to the Vindhya Mountains and the river Nermada, which separate the Dekhan from Hindústan.

3. A variety of mountain-tribes in the Dekhan, with regard to whom it has not been proved whether they are of one original stock with the proper Tamuls, or tribes really distinct from that people, expelled by them into mountainous and inaccessible tracts.

4. A greater number of petty barbarous tribes between the Indian and the Indo-Chinese Peninsulas; that is, in the countries not far distant from the lower course of the great river Brahmaputra. The vicinity of so many different tribes to this great channel, joined to the fact that many of the nations described have considerable resemblance to the races of High Asia, render it not improbable that they may have descended in remote ages along the course of rivers from the countries lying northward of the Himalayan border, and may have taken up their abode in the valleys, and even on the mountains, lying near the channel.

### 1. *The Singhalese Race.*

The Singhalese are the inhabitants of the interior and southern part of the great Island of Ceylon, the Selendiva of ancient geographers, called also Taprobane, probably from Tambapanni, one of its Indian names.* The Singhalese race occupies about one half of the whole island, from Chilaw to Batticaloa. The people termed particularly Singhalese are the inhabitants of the countries near the

* The Sanskrit name is thought to have been Tamravarna, copper-coloured, from the colour of the soil.—Ed.

southern coast; in the interior are the Kandians. Besides these, there is a tribe of wild people in the interior, who inhabit the mountainous tracts in the neighbourhood of Batticaloa, termed Vaidas, or Vaddahs, who exist in the most savage state, and support themselves on the spontaneous fruits of the earth, and on the prey which chance brings within their grasp. It has been supposed by some that the Vaddahs are a distinct race from the Singhalese and the aborigines of the island. They may be, and probably are, among the oldest inhabitants of it; but it was long ago known to Knox that the Vaddahs speak a dialect of the same language with the Singhalese; and they are therefore probably descendants from one original stock, of which some tribes became civilised, while others remained in their original barbarism. The Kandians and the Singhalese resemble each other in manners, language, and religion, and it is evident that they were originally one people. They are all worshippers of Buddha, whose religion was introduced into Ceylon some centuries before the Christian era, and established by Asoka, king of Magadha, who reigned over a great part of India soon after the invasion of that country by Alexander the Great. The Vaddahs, according to Mr. Cordiner, are still of the Hindú religion, which prevailed, before the introduction of Buddhism, through the whole island, where pilgrims who now stop at the Isle of Ramisseram formerly continued their progress to the Temple of Siva, at Divinúr, in the southern extremity of Ceylon. The Hindú religion still prevails among the Tamuls, or Malabars, who inhabit the northern part of the island.

The following account of the Singhalese is given by Dr. Davy, the latest and best informed writer who has described the Island of Ceylon and its inhabitants:—

"The pure Singhalese of the interior, whom alone I shall describe, are completely Indians in person, language, manners, customs, religion, and government.

"Like Indians in general, the Singhalese differ from

Europeans less in features than in the more trifling circumstances of colour, size, and form. The colour of their skin varies from light brown to black. The colour, too, of their hair and eyes varies, but not so often as that of the skin: black hair and eyes are most common; hazel eyes are less uncommon than brown hair; grey eyes and red hair are still more uncommon; and the light blue or red eye and light flaxen hair of the albino are the most uncommon of all. In size they generally exceed the lowland Singhalese and most of the natives of the coast of Coromandel and Malabar: they are inferior to Europeans. Their average height may be about five feet four or five inches. They are clean made, with neat muscle and small bone. For Indians they are stout, and generally have capacious chests and broad shoulders, particularly the inhabitants of the mountainous districts, who, like Highlanders in general, have rather short, but strong and very muscular, thighs and legs. Their hands and feet are commonly very small; indeed, so much smaller than ours, that they appear out of proportion. The form of their head is generally good, perhaps longer than the European, —a peculiarity, according to Dr. Spurzheim, of the Asiatic. Their features are commonly neat, and rather handsome: their countenances are intelligent and animated. Nature has given them a liberal supply of hair, which they universally allow to grow on their face, as well as head, to a considerable length, being of opinion that the beard does not deform but improve the face; and certainly, in many instances, I have seen it have the effect of giving to the countenance an air of dignity that would have disappeared with the use of the razor.

"The Singhalese women are generally well made and well-looking, and often handsome. Their countrymen, who are great connoisseurs of the charms of the sex, and who have books on the subject and rules to aid the judgment, would not allow a woman to be a perfect belle, unless of the following character, the particulars of which I shall give in detail as they were enumerated to me by a Kandian courtier, well versed and deeply read in such matters:— 'Her hair should be voluminous, like the tail of a peacock, long, reaching to the knees, and terminating in graceful curls; her eyebrows should resemble the rainbow; her eyes the blue sapphire, and the petals of the blue manilla flower; her nose should be like the bill of the hawk; her lips should be bright and red, like coral or the young leaf of the iron-tree; her teeth should be small, regular, and closely set, and like jessamine buds; her neck should be large and round, resembling the berrigodia; her chest should be capacious; her breasts firm and conical, like the yellow cocoa-nut; and her waist small, almost small enough to be clasped by the hand; her hips should be wide; her limbs tapering; the soles of her feet without any hollow; and the surface of her body in general soft, delicate, smooth, and rounded, without the asperities of projecting bones and sinews.' "*

Dr. Davy has, in another work, given us a description of three individuals whom he had seen of the race of Vaddahs, or Vaidas.† "They belonged to a large party who had come to Kandy with a tribute of dried deer's flesh and wild honey. They were quite naked, with the exception of a scrap of cloth. The hair of their head and beard was long and matted, and had never been cut or combed; their eyes were lively, wild, and restless; they were well made and muscular, but of a spare habit; and, in person, they chiefly differed from the Kandians in the slightness of their limbs, the wildness of their looks, and their savage appearance. According to their own account of themselves, they came from the neighbourhood of the Lake of Birtenne, where they subsisted on game which they killed in the chase, some roots, and wild fruits, and a little grain of their own growing. They were profoundly

* "Account of the Island of Ceylon," by John Davy, M.D. F.R.S. &c.

† "Researches, Physiological and Anatomical," by John Davy, M.D. F.R.S. London, 1839. 2 vols. Vol. ii. p. 177.

ignorant, could not count above five, were hardly acquainted with the rudiments of any art, and, though they feared demons, as they did wild beasts, they had no knowledge whatever of a supreme beneficent Being, and not the slightest notion of any state of existence after the present. Yet these men considered themselves civilised, in comparison with the wilder tribes of Vaidas, who never leave their sylvan haunts, and whom I have heard Kandians of a bordering province describe as living almost entirely on raw animal food, as going quite naked, as having no superstition, and, in fact, as being in a state very little removed from that of brutes."

It has often been observed that albinos are frequently seen in Ceylon. Dr. Davy speaks of such persons; I shall describe his remark on one of them. He says, "The young albino, twelve years of age, in England, and certainly in Norway, would not be considered peculiar; for her eyes were light blue, and not particularly weak; her hair of the colour that usually accompanies such eyes; and her complexion fresh and rather rosy. She had considerable pretensions to beauty, and was not without admirers among her countrymen. It is easy to conceive that an accidental variety of the kind might propagate, and that the white race of mankind is sprung from such an accidental variety. The Indians are of this opinion, and there is a tradition or story amongst them in which this origin is assigned to us."

## 2. *The Tamulian Race.*

The proper Tamuls are the inhabitants of the northern part of Ceylon and of the southern portion of the Dekhan. Their language and race are spread over the country from Cape Comorin on the coast of Coromandel as far northward as Pulicat, and over the greater part of Barramahal, Salem, and Coimbatore. To the westward they border on the Malayalma language and the Malabars, who speak a dialect of the same idiom, as do the inhabitants of the western coast of the Peninsula as far as the extent of Tuluva. All those people may be considered as belonging to the Tamulian nation in a stricter sense than that in which I have used the term Tamulian race.

To the Tamulian race I refer other great nations in India whose idioms are sister languages of the people of the Tamul country. These are the people of Telingána on the eastern side of the Dekhan, the kingdom of Andhra of Sanskrit authors, whose idiom is the Telinga, or Telŭgŭ; secondly, the Karnatas, or Canarese, who inhabit the table-land above the Ghauts in the interior of the Peninsula and the country of Mysore. The people of the district of Canara on the west, who speak the Tuluva dialect, the Karnatas in the interior, and the Telingas on the eastern side of the Peninsula, are the most northern nations which belong to the Tamulian race. The Mahrattas to the northward, and the Uriyas, or people of Orissa, speak dialects or *bháshás* of the Sanskrit, and are of Hindú extraction.*

[The physical characters of the Tamulian race are best described by J. R. Logan, Esq.†. After noticing the mixed features everywhere apparent in the man of India, he proceeds to show that the original type, as least exposed to foreign interference, must be looked for in the south. He then goes on to say: "The diversity even here, is so great, as to show that there has been much mixture; but there are certain widely prevalent characters, most of which are not Arian nor Tibetan, and are even distinct from Ultra-

* It is a question whether all the languages of India are not of one origin, with the only difference that the so-called Sanskrit dialects have received a much more copious infusion of Sanskrit words than the southern tongues. The Editor is of this opinion.—See the valuable paper of Dr. Stevenson in the Journal of the Bombay Branch of the Royal Asiatic Society, vol. i. p. 103; and the very valuable essays of J. R. Logan, Esq., in the Journal of the Indian Archipelago, printed at Singapore,—a periodical not sufficiently known in Europe.—ED.

† See Journal of the Eastern Archipelago, vol. vii. p. 302. Singapore 8vo. 1853.—ED.

indian. The more important of these characters are a pointed, and frequently hooked, pyramidal nose, with conspicuous nares, more long than round; a marked sinking in of the orbital line, producing a strongly defined orbital ridge; eyes brilliant, and varying from small to middle-sized; mouth large, lips thick and frequently turgid; lower jaw not heavy, its lateral expansion greater than in the Arian, and less than in the Turanian type, giving to the middle part of the face a marked development and breadth, and to the general contour an obtuse oval shape, somewhat bulging at the sides; forehead well formed but receding, inclining to flattish and seldom high; occiput somewhat projecting; hair fine; beard considerable, and often strong; colour of skin very dark, frequently approaching to black."]

The literature, arts, religion, and peculiar civilisation of the Dekhan, are Indian or Brahmanical; and all the languages of the Tamulian nations above mentioned, though fundamentally different, have derived great additions from the Sanskrit. It has been a question whether the nations of the Peninsula had any culture of their own previously to their subjugation under the conquerors or priests of Hindústan; and some suppose that they were till that period on a par with the tribes of the mountains and forests in the interior, who have fled from the approach of civilisation, and preserve in the remote and least accessible parts of the country their pristine barbarism. The earliest poetical composition in the Sanskrit language represents the inhabitants of the Dekhan in this point of view. The celebrated "Ramayana," the oldest epic poem of the Hindús, which is allowed on all hands to be more ancient by many centuries than the "Iliad,*" has for its argument a war said to have been carried on by the hero

* The age of the Ramayana is now generally admitted, even by the most decided partisans of the antiquity of Sanskrit literature, to be not older than the fourth century B.C. See Wilson's "Rig-Veda" Sanhita, Introd. p. xlvii. London, 1850.—ED.

Ráma, king of Ayodhya or Oude, in northern Hindústan, against Ravana, king of Lankadwipa, or Ceylon, who possessed a great part of the Peninsula. The aim of Ráma's exploits in this southern region, where no mention occurs in the poem of inhabitants of towns or even of forests and caves, except hermits, apes, bears, vultures, demons, and magicians, was to deliver holy penitents from the fear of Ravana and his giants, who possessed Ceylon and the Dekhan. At the head of these penitents and pilgrims was the Muni Agastya, the celebrated apostle of the religion of Siva, whose efforts were seconded by Rama and his followers. At what period the Brahmans, and the warlike Kshatriyas who assisted them, really succeeded in establishing their dominion in Ceylon, is unknown, but it must have been at an early period; since Buddhism, which superseded the region of the Vedas for some centuries in the Dekhan, and finally in the island, was, as we have said, established in those countries by Asoka, who is known to have been contemporary with the first Antiochus.

Those who have devoted most study to the history of India are yet of opinion that a peculiar civilisation, and even the art of writing, existed in the Tamulian countries prior to the conquest of the Hindús; but its sources are wholly unknown, and its character can only be matter of conjecture. The earliest commerce with the western region of the world probably took place subsequently to the Hindú conquest.*

### 3. *The Parbatíya, or Mountaineers, or Wild Tribes of India.*

The Sanskrit name of Parbatíya, or पर्व्वतीय, is given to several races who inhabit hilly countries in

* Among the articles of earliest traffic, as Professor Karl Ritter observes, was probably tin; and it is likely the Greeks obtained this metal from the East, in the age of Homer. Kastira is the Sanskrit name for tin, whence the Greek κασσίτερος. This metal had obtained its Sanskrit name in India, prior to the *Tooje era.* [This is unintelligible to the Editor; probably the "Trojan war" was meant, the mistake being typographical only.—Ed.]

northern India. The meaning of the term is mountaineer, and it may well be generalised and used as a common name for all the tribes who live remote from cities and cultivated countries, and maintain a savage existence amidst woods and forests. In this sense there are many Parbatíya races in different parts of Hindústan and the Dekhan. Ethnology is in far too imperfect a state to render it possible at present to determine what relation these tribes bear to each other, and to the civilised nations who are nearest to them. There is reason for believing that some of the wild races in the Dekhan are allied to the Tamulian tribes; and it is not improbable that most of them are descended from people of that stock who refused to receive the apostles of the Hindú theology, and of civilisation and slavery. They are, however, now very different from each other in different parts of India, both in moral and, more especially, in physical characters; some being vigorous and finely formed, others diminutive and puny. The difference may be explained in many instances by reference to the climate and local influences under which the several tribes exist.

Some progress has been made almost yearly in extending our knowledge of the native races of India, and particularly of the wild mountain tribes of the Dekhan. The names of these tribes are almost innumerable: they are often derived from accidental distinctions, and from districts of the country where they are found, and different branches of the same people are often to be recognised under various designations. The Bhíls, the Kúlis, the Ramúsis, the Wáralis, and the Katodis, are the most celebrated tribes in the north-western parts of the Dekhan, where they reach from the Vindhya chain of mountains to the forests of northern Kankana, or Concan. In the north-western and more central parts are the Gonds, in Gondwana; the Pulindas, or wild tribes of Orissa, consisting of the Kolis divided into thirteen tribes, each having a different name; the Khonds to the southward of the river Mahanadi; the Sours towards the forests of Katak; the Yanadu-Yati, or wild people of Sri-hari-cotta, among whom the Coy-Vandlu are particularly mentioned. In the Nilagiri, or Neilgherry hills, in the southern part of the Dekhan, towards the junction of the two chains of Ghauts, are various barbarous tribes, termed by Mr. Hough, who has described them, Thodaurs, Buddagurs, Curumbars, and Kothars. To these must be added the Cohatars, who occupy the summits of hills.* In physical character these races differ greatly among themselves: some of them are small, shrivelled, black savages, who have been thought to resemble the Negroes of Africa;† others are tall and athletic, handsome, with features resembling the European type. These are the inhabitants of the elevated tracts, where a tolerably cool and salubrious climate exists: the blackest and most diminutive tribes are found in the jungle near the rivers, and in low unhealthy districts. Yet it is probable that most of these tribes are of one aboriginal race, since, where vocabularies of their various dialects have been collected, they have been found generally to bear some traces of affinity to the Tamulian, or its sister languages; that is, to the idioms of the civilised nations of the Dekhan. We have, however, as yet scarcely sufficient information to sanction a positive conclusion on this subject.‡

* The reader will find such information as I have been able to collect respecting the history and physical and moral characteristics of these different tribes, in the fourth volume of my "Researches."

† The women of the Khonds are described as ugly, with large mouths and projecting lips, flat and broad noses, and high cheek-bones.

‡ Grammars and vocabularies of these languages are beginning to appear, several of them in the pages of the Indian Scientific Journals, and others, such as the Rev. J. Phillips's Grammar of the Sántál,—a dialect spoken by a very wide-spread tribe of the Kole family (Calcutta, 1852),—as independent works. A very valuable summary of these publications, with many collations from his own extensive knowledge, is given by J. R. Logan, Esq., in the pages of the Singapore Journal, shewing the Tamulian analogies of these languages as well as their connection with Tibetan and Ultragangetic tongues. See especially the seventh volume of that periodical.—Ed.

Several of these mountain tribes, or Parbatíyas, have acquired some tincture of the Hindu religion from Brahmans who have been among them; but this is every where very slight, and not enough to produce any remarkable change in their habits. They have, for the most part, retained their barbarous idolatry, which is to be found in almost all nations in the lowest stage of rudeness and ignorance. Some of them practise human sacrifices, and bury their victims, who are generally of a different race, alive. The oath administered to the Khonds is characteristic. It is as follows: "Oh, Father! I swear, and if I swear falsely may I become shrivelled and dry like a blood-sucker, and die—may I be killed and eaten by a tiger—may I crumble away like the dust of this ant-hill—may I be blown up like this feather—may I be extinguished like this lamp." An appropriate action accompanies each article of this barbarous oath, if oath it may be called.*

The finest race among all these different tribes are the Tudas, of the Nilagiri hills. They are the race called Thodaurs by Mr. Hough. They have been described in a particular memoir by Captain Harkness; and several portraits of individual Tudas, beautifully drawn, which have been placed in the museum of the Royal Asiatic Society, afford specimens of their features and countenance. Copies of two of these portraits, by the kind permission of the Society, have been taken for the illustration of this work, as may be seen in an adjoining page. The Tudas are an interesting people: they seem to be susceptible of culture, and may hereafter become a civilised and powerful nation. For a race of mountaineers they appear to be

* Some valuable papers of much interest, by Captain Macpherson, entering with minuteness into the religious observances of these people, and describing the circumstances and ceremonies which accompany their numerous sacrifices of human victims, are printed in the Journal of the Royal Asiatic Society. See vol. vii. p. 172, and vol. xiii. p. 216.—ED.

remarkably gentle and intelligent. Their appearance, according to Captain Harkness, is very prepossessing: they are generally above the common height, athletic, and well made; their bold expressive countenances declare them to be of a different race from their neighbours. They are bare-headed, wear their hair six or seven inches long, parted from the crown into bushy ringlets, and of a jet black,—their beards likewise grow; a large, full, expressive eye, a Roman nose, and pleasing contour, gravity varying into cheerfulness, distinguish them from other natives of India. They wear rings of silver and gold,—wear only a light, short garment. The women are somewhat fairer than the men, have the same expressive features, but with a feminine cast, and wear long tresses of black hair.

They are a pastoral people, have no towns or villages, but live in solitary houses, and feed sheep and cattle.

Their language is considered as distinct. It has, *perhaps*, some resemblance to the vernacular idioms of the Peninsula, and principally to the Tamil. There are two sounds, the *Zha* and *Ukh*, which are of constant occurrence in the Tuda, and which, among the idioms of the low country, are peculiar to the Tamil, and its sister dialect the Malayálma. The pronouns, the plural, the honorary terminations of the verbs, are nearest to the Tamil. With these exceptions it differs widely from the Tamil, and all other known languages. It is merely an oral dialect.*

* The Tuda language has been successfully investigated since the above was written, and clearly shewn to be a dialect of the Dekhanic languages. The following extract from a paper by D. B. Schmid, a six years' resident among these people, and well acquainted with several Indian tongues, is decisive on this point, as well as on the idiom of the Budaghers,—the Buddagurs of the preceding page. "The Todaver language is a genuine but very rude dialect of the ancient Tamul, the words of which are in many cases so greatly changed, but changed according to certain rules, that only a deeper study enabled me to recognise the identity of both languages; and the comparison of these Todaver words with the Budagher and Canarese words shows to evidence that the Tamul, Todaver, Budagher, and Canarese languages are links of a closely connected and unbroken chain of one original language, and that the Todaver dialect is by far more closely connected

On the mountains the Tudas have erected temples, consisting of large stones. In some of these are numerous urns and figures of animals, such as the buffalo, the tiger, the peacock, and the antelope. Their religion has no resemblance to those of the Buddhists, the Moslims, or any other people. They salute the sun on its rising, and believe that after death their souls go to Huma-norr, or Om-norr, a country respecting which they ask for information.

In northern and proper Hindústan, and towards the eastern part, a remarkable people are the natives of the Rajamahal hills. Their physical characters and manners are peculiar, but they are said not to be so savage as the Bhíls and Gonds: they have a religion and a priesthood of their own, which are said to be entirely different from those of the Hindús. Their language is said to bear a resemblance in some words to the Tamul, as it plainly does to the idioms of some tribes beyond the Brahmaputra.

To these barbarian races within the boundaries of India generally so termed, we must add a variety of tribes inhabiting different tracts of no great extent in the valley of the Brahmaputra, and in the countries near the mouth and lower course of that river, and the borders of the Bay of Bengal. Among these are the Ahoms, the Garros, the Cachars, the Cossyahs, the Manipurs, Miris, Abors, Kangtis, and the Nagas or Kukis. Among these races a great variety is perceived in physical characters: some approximate to the physiognomy of the Hindús, others to that of the Bhotiyas.*

---

with the Tamul than with the Canarese" (Journal of the Bombay Branch of the Royal Asiatic Society, vol. iii. p. 52).—Ed.

* The reader who is curious for such information will find all that I have been able to collect in illustration of the history of these races of people, in the fourth volume of my "Researches into the Physical History of Mankind," pp. 219-228.—Dr. Prichard.

A valuable account of the Nagas is given in the Bengal Journal for April 1849, by W. Robinson, Esq., whose position as Inspector of Government Schools in Asam gave him ample opportunities of being acquainted

If we were at liberty to conjecture the way by which the different races of India first peopled it, and the quarters whence they originated, from the feeble indications of analogy in physical traits and language as yet known, it would seem most probable that as the Indo-Chinese tribes descended from the high countries of Yunnan and Laos, along the shores of the Mekon, the Menam, the Saluen, and the Irawadi, into the maritime and southern parts of the Eastern peninsula, so the Allophylian tribes in Hindústan and the Dekhan descended from the north-east along the Brahmaputra, and finding no obstacle to their progress, spread themselves over the plains of central Hindústan and the parts of the Peninsula above and below the Ghauts, till they arrived in the Island of Ceylon. They had probably occupied all these countries before the time when the Hindús, of Arian or Indo-European descent, crossed the Indus. By them the ori-

with the people of the country. Speaking of the Nagas, he says they are "known to the Asamese by the general name of Nagas. Whatever may be the origin of the word, it appears that the appellation is entirely unknown to any of the Hill tribes themselves. They are divided into numerous communities or races, and they know themselves by the designation of their respective tribes only, and not by any name common to all the races.

"There appear, however, to be some marks by which these tribes are distinguished from their neighbours, and even common ties by which they are all bound together as one people, though at present divided into tribes by a diversity of dialects. These dialects are sometimes so different that two adjoining tribes cannot converse together, except through the medium of a third dialect common to both; yet they are said to intermarry, and form connections and alliances with each other, which they do not do with tribes not belonging to the Naga community.

"The Nagas also appear in general to be distinguished from their neighbours by physical conformation, for though there is much difference in this respect among them, yet they are in common remarkable for extremely coarse savage countenances, and dull, timid, heavy dispositions."

The Hill country of Assam, according to Mr. Robinson, is bounded on the west by the Kopili River, the south bend of the Barak, and the eastern frontier of Tipperah, in E. long. 83°; on the north, by the valley of Asam; on the east and south-east, by the mountains dividing it from the Borkhamti country, in W. long. 97°; and on the South, by the 23d parallel of latitude. —Ed.

ginal inhabitants were expelled from Hindústan, where they left only a few barbarous hordes in the hilly tracts of the eastern side. In the Dekhan they maintained their independence much longer; and the population of that country, as well as of Ceylon, is still in great part of the aboriginal stock.

[A valuable memoir by General Briggs in the 13th volume of the Royal Asiatic Society's Transactions, gives a full account of the numerous aboriginal tribes of India, who are, as he with every probability concludes from physiological and philological evidence, all of one race; having languages and habits allied to those of the Scyths or Tartars of Asia, and distinct from and opposed to those of their Sanskrit-speaking conquerors. Large tracts of India are inhabited solely by this race, and evidence abounds of their former, and in some cases even of their recent power, such as may be seen in Gondwana, a region of 70,000 square miles, "containing a vast population under their own chiefs, and retaining their primitive habits, physiognomy, and religion, unmixed by connexion with the Hindús." Of their most prominently distinct habits he recounts their entire disregard of caste, their eating of all kinds of flesh without distinction, their use of fermented liquors, the bloody sacrifices which accompany their religious rites, their patriarchal institutions, their marriage with widows, all utterly abhorrent to present Hindú habits, although traces of some of these practices are found in the most ancient writings. The General further says, "the aborigine is not more distinguished in his other habits than he is in his moral virtues from the Hindús. The man of the ancient race scorns an untruth, and seldom denies the commission even of a crime that he may have perpetrated, though it lead to death. He is true to his promise, hospitable and faithful to his guest, devoted to his superiors, and is always ready to sacrifice his own life in the service of his chief."]

## CHAPTER X.

### REMAINS OF ABORIGINAL NATIONS PRESERVED IN MOUNTAINOUS TRACTS OF THE WEST.

We have seen that the extreme borders of Europe and Asia, towards the north, contain the remains of tribes who were probably more widely spread before the advance of the Indo-European nations from the south. We must now observe, that in the midst of regions long ago conquered by the Arian and Syro-Arabian races, there are mountainous tracts, difficult of access, where remains are still to be found of a more ancient stock of inhabitants. Thus the chain of Caucasus harbours at this day many tribes whose origin is unknown, and who are proved by their language to have no affinity with the inhabitants of adjoining countries.* In the western part of Europe, the Pyrenees and parts of the Alpine chain were inhabited at the Christian era by tribes, perhaps aboriginal, of Iberians, Ligurians, and Rhætians: of these the former have preserved to this day, in the mountains between France and Spain, their peculiar speech, which appears to be a relic of the language once spoken over a great part of western Europe, from Sicily to the Straits of Hercules. Again in the north of Africa, the chain of Mount Atlas, and other tracts in the interior, are inhabited by remains of the ancient Lybians. The Lybian race occupied the southern coast of the Mediterranean, and, jointly with Iberian tribes, peopled several islands in that sea. In this country the Phœnicians founded colonies, and spread far and wide the Syro-Arabian language. The Phœnician language was already the idiom of civilised Africans, as we learn from the researches of Gesenius, in the days of Massinissa and Jugurtha; and, after the later incursions

* The affinities of these languages, as shown by Dr. R. G. Latham and B. H. Hodgson, Esq., appear to be with those of the people of the Himalaya mountains and the country beyond the Ganges (see also note in the next page).—Ed.

of the Arabs, the aboriginal Lybian was entirely banished into the desert and mountainous parts of the interior.

These nations have been preserved from extinction by similar circumstances. I shall briefly touch upon their history in one section, though I do not mean to infer any family connexion between them. This, indeed, has been conjectured; it has been supposed from the identity of their name, that the Iberi at the foot of Mount Caucasus may have been the original stock whence came the Iberians of ancient Spain; but the entire difference of the language and habits of these nations has proved this suggestion to be erroneous; nor has the conjecture long ago thrown out, that the primitive inhabitants of the peninsula were of African origin, and akin to the people of Mount Atlas,* obtained the least support from philological researches.

### 1. *Caucasian Nations.*

The high chain of Mount Caucasus, which has in many ages of the world formed a bulwark or barrier, defending the more civilised or luxurious inhabitants of Southern Asia from the barbarians of the north, contains within itself places of difficult access, and easily defensible against powerful invading armies. In these tracts many rude tribes maintained their independence against Persians, Greeks, Romans, Mongolians, and Turks, and are still but imperfectly subdued by the armies of Russia. The inhabitants of this region belong to ancient races, different in language and manners from all other nations; and among them are several tribes equally different from each other.† From this remark we must exclude the Ossetes, or Ossetines, already mentioned, who speak an

* This notion was supported by Jezreel Jones, author of a dissertation inserted in Chamberlane's "Oratio Dominica."

† It must, however, be observed, that marks of ancient connexion, more or less remote, between some of the Caucasian languages and the dialects of northern Siberia, are strongly indicated.

Arian or Indo-European dialect, and are thought to be remains of the German Alani:* they live near the sources of the river Terek, and are supposed to be about forty thousand souls.†

The Basian tribes must likewise be excepted, who with the Chumyks live in the interior of the Caucasus; they, as well as some Turkoman tribes between the Kuma and the Terek, are of Turkish origin. The proper or aboriginal inhabitants of the Caucasian region are known, after a long and careful research into their history and languages,‡ to consist of four distinct races, in each of which are several tribes unintelligible to each other. They are distinguished by their local situation, as the western, middle, eastern, and southern Caucasians; the latter division including the Georgians, whose principal abodes were on the river Kúr, or Cyrus. I must enumerate the principal tribes belonging to each of these families.

1. The Western Caucasian division includes two nations long supposed to be distinct, but proved by a careful examination of their languages to be branches of one stock, though they differ in physical characters. These nations are the celebrated Circassians and the Abassians.

The Abassians appear to be very ancient inhabitants of the north-western part of the chain of Caucasus. They are chiefly pastoral and predatory people. They are divided into two nations, termed the Great and the Little Abhasa.

The Abassians, according to Klaproth, are distinguished

* The term Alan appears to be rather topographical than ethnological: it signified "mountaineer," and has been borne by people of very distinct national families. See Ammianus Marcellinus, "Alani ex montium adpellatione cognominati," quoted by Neumann in "Die Völker des südlichen Russlands," p. 36.—Ed.

† "Russland und die Tscherkessen," von K. F. Neumann. Stuttgart, 1840.

‡ Dr. Guldenstädt's "Beschreibung der Kaukasischen Länder," von J. Klaproth. Berlin, 1834. Klaproth's "Asia Polyglotta," p. 109. Klaproth's "Sprachatlas."

by narrow faces, heads compressed at the sides; by the shortness of the lower part of their faces; by prominent noses, and dark brown hair.

The Circassians are to the eastward of the Abassians; they inhabit the country between Caucasus and the Kuban, and, farther eastward, the provinces of Great and Little Kabarda, on the Terek. The people of these provinces are also called Kabardines. The Circassians term themselves Adigi, and by the Tartars are named Tscherkess; whence the appellation by which they are known in Europe. Their country abounds with high forest tracts, and their climate is cold.

Pallas informs us that the Circassians are a handsome race of people. "The men," he says, "especially among the higher classes, are mostly of a tall stature, thin form, but Herculean structure. They are very slender about the loins, have small feet, and uncommon strength in their arms. They possess in general a truly Roman and martial appearance. The women are not uniformly Circassian beauties, but are for the most part well formed, have a white skin, dark brown or black hair, and regular features." He adds, "I have met with a greater number of beauties among them than in any other unpolished nation." Other travellers represent a mixture of red in their hair as a characteristic of the Circassians.* Klaproth says, "They have brown hair and eyes, long faces, thin, straight noses, and elegant forms."† Reineggs denies their claim to superior beauty. He says, "I know not what can have given occasion to the generally received prejudice in favour of the female Cherkessians. A short leg, a small foot, and *glaring red* hair, constitute a Cherkessian beauty."‡

2. The Middle Caucasians inhabit the high country above Kabarda and the habitations of the Circassians, towards

* Pallas's "Travels in Southern Provinces of the Russian Empire."

† Klaproth's "Travels in the Caucasian Countries."

‡ Reineggs' "Allgemeine historische-topographische Beschreibung des Kaukasus," &c.

the sources of the Terek and other rivers which flow into it from the heights of Caucasus. There are several tribes or nations belonging to this division of people, who, however, seem to be nearly allied in language. The principal of these are the Mizjeghi, a name which, according to Klaproth, comprehends the Ingushi, and the Chechenghi, or Tashi; the latter being to the southward, nearer to Georgia. The Ingushi are termed by Guldenstädt, Kisti.

3. The Eastern Caucasians, or the Lesghi, are the inhabitants of the eastern region of Caucasus, which is termed Lesghistan. They are divided by Guldenstädt into seven tribes or nations, one of which is the Avars, supposed to be the remains of the people once so formidable under that name.

4. Several nations in the southern tracts of the Caucasus belong to the Georgian or Grusian race, whose country is termed by the Persians Gúrjistan, from the river Kúr, whence its European name of Georgian. The Proper Georgians are the Kart'huli; or inhabitants of Kartuel and Imeretia. The old Georgian language, into which the version of the Scriptures was made, is an obsolete dialect of the Kartuel. The modern dialect of this province is the Georgian language, properly so termed.

A second division of the Georgian race are the Mingrelians, inhabitants of the ancient Colchis, the modern Mingrelia and Guriel, on the banks of the Black Sea. The Soani, a tribe of mountaineers in the southern Alpine tracts of Caucasus, are another tribe of the same race. A fourth are the Lazians, a people well known in the middle ages as a barbarous and predatory tribe on the sea-coast of the Euxine, as far to the westward as Trebisond.

The Georgians are a people of European features and form. Reineggs says that their women are more beautiful than the Circassians, but that the prevalent complexion of the race is not so fair as that of the Circassians who are natives of the higher country of Caucasus.

## 2. *Aboriginal Inhabitants of the Coasts and Islands of the Mediterranean.*

If we may give credit to obscure traditions collected in various quarters by ancient historians and geographers, among whom are Thucydides, Strabo, and Pliny, the islands of the Western Mediterranean and the coasts of both sides of that sea were occupied, in the earliest periods to which history can reach back, by tribes of two different races, termed Lybians and Iberians, who were often intermixed in the same island, or shared between them its possession.

1. The language of the ancient Iberians has survived to our times in the vernacular speech of the Biscayans in Spain and the Basques of France, inhabitants of mountainous districts in the confines of the two countries. The national appellation of these people in their own idiom is Euskaldunes; and they term their language the Euskara, or Euskarian speech. They were formerly divided into a great number of tribes. The present Euskaldunes are descendants of the Vascones and the Varduli: these tribes occupied the country which the people who speak the Euskarian language now possess. It is only by mistake, or through national vanity, that they have been termed, or have termed themselves in modern times, Cantabrians; the Cantabri lived in a part of Spain where the Biscayan is not spoken.* It has been proved that the Euskarian is quite distinct from the Celtic and from all other Indo-European languages, and that in structure it most nearly resembles the idioms of the native tribes of America. Yet this resemblance hardly amounts to a family relation, or to that kind of connexion which proves a common origin, the difficulties which lie against such an inference being taken into the account.

The ancient Iberians were a people early cultivated,

* "Prüfung der Untersuchungen über die Urbewohner Hispaniens," von Wilhelm von Humboldt. 4to. Berlin.

and they had the use of letters derived originally from the Phœnicians, and nearly resembling some of the alphabets of the old Italic nations. They are first known in history as inhabitants of the northern coast and islands of the Mediterranean. The earliest, not fabulous, people of Sicily were of this race; and the researches of William von Humboldt seem to have proved that vestiges of their idiom are still to be traced through a considerable part of Italy, where they, perhaps, preceded the Italic nations of the Arian race. The coast of Gaul, from the Rhone westward, was occupied by Iberians, who there lived intermixed with Ligurians, the latter people having sole possession of the maritime tracts between the Rhone and Italy. Such is the account given in the Periplus of Scylax, which Niebuhr supposed to have been a compilation from the nautical records of very ancient voyagers. The Ligurians are said to have come from the neighbourhood of the river Ligys or Ligyros, which has been supposed to be the Loire, and to have expelled the Iberians from a part of their ancient territory. These events were probably prior to the entrance of the Celts into Western Europe. The latter people, who were more warlike than the Iberians, appear to have dispossessed them of a considerable part of Spain; for traces of Celtic habitation in the names of towns and tribes have been discovered by Humboldt through nearly all the western half of the peninsula. The Pyrenees, however, always remained in the possession of the Iberians. Iberians, also, were among the early inhabitants of Corsica, Sardinia, and the Balearic Isles, where they bore the name of Balari. In several instances, Iberians and Lybians were found in the same island.

The modern people who speak the "Lengua Bascongada," or Bascuence, in the Spanish provinces of Biscay and Guipuzcoa, and the Basque or Labourdine dialect in the Basque provinces of France, are a race very interesting by their peculiar character and habits. A late writer has given a lively description of them, from which

I shall cite some passages that may answer the purpose of an ethnographical sketch:*—

"Les Basques, tels qu'ils existent maintenant, hommes et femmes, ont dans leur extérieur un attrait que l'on ne trouve chez aucune autre race humaine. Les hommes, de taille moyenne mais bien proportionnée, portent l'expression de la vigueur et de l'agilité: le proverbe *courir comme un Basque* est parfaitement juste. Leur costume léger adapté à leur climat et à leurs mœurs, laisse apercevoir tous leurs mouvemens, naturellement plus gracieux que ceux d'aucun autre peuple. Une jaquette brune négligemment jetée sur l'épaule gauche, un gilet rouge ouvert, une chemise toujours très propre, une culotte collante assujétie au moyen d'une écharpe rouge, des bas bleus ou bruns, d'élégantes sandales faites avec des nattes de chanvre et attachées avec des rubans rouges, forment l'habillement du jeune Basque: sa tête est couverte d'un petit bonnet plat qu'il met de côté, ou bien ses cheveux sont retenus dans un réseau de soie. D'ordinaire le Basque ne porte point d'armes: mais il quitte rarément son bâton ferré, dont, en cas de rixe, il sait se servir avec une merveilleuse adresse, tant pour l'attaque que pour la défense. Il est difficile de donner de la beauté et des graces des femmes une idée qui approche seulement de la réalité. Que dire d'elles, si ce n'est qu'elles ont les plus belles proportions, la taille la plus fine, le teint le plus ravissant; que leurs beaux bras, leurs belles mains, leur pied mignon sont dans la plus parfaite harmonie avec un profil vraiment Grec! Mais il est impossible de faire comprendre combien tous ces attraits sont relevés par l'ineffable aménité de tous leurs mouvemens, par la grace de leur démarche, par le sourire malin qui voltige autour de leur bouche de corail, et embrase d'un nouveau feu leurs yeux noirs déjà si vifs; de décrire avec quelle adresse elles lancent en l'air leur fuseau ou maintiennent en équilibre sur leur tête une

* "Extrait du Voyage de Lunemann dans les Pyrenées." "Nouvelles Annales des Voyages." Paris, 1831.

petite cruche de terre: avec quel art elles nouent autour de leurs cheveux le mouchoir dont les longs bouts retombent sur leur nuque; de faire sentir à quel point leur va bien leur chapeau de feutre blanc, leur fichu d'un rouge éclatant, et leur jupon court écarlate; de peindre le feu de leur regard, le pantomime de tous leurs membres qui accompagne leurs discours. On reproche ordinairement à ces créatures enchanteresses un peu de coquetterie et beaucoup de légèreté; mais j'ai tout lieu de croire qu'en les en accusant, on prend l'apparence pour la réalité: le ton mièvre, par exemple, avec lequel elles crient à tous les étrangers leur *Lgun hon Jauna*, la manière folâtre dont elles les accostent et les turlupinent; leur disposition constante à rire et à plaisanter,—peuvent très bien prendre leur source dans leur naïve candeur même, et passer à tort pour de la facilité. Du moins cette prétendue facilité ne me paraît-elle guère compatible avec cet esprit profondément religieux, avec cette innocence d'expression, avec cette réserve dans toutes les actions, avec cette modestie dans tous les mouvemens que j'ai trouvé à Ustarritz, à Hasparren, et dans d'autres bourgs éloignés. Au reste, la sévère retenue que les hommes observent en leur présence, et qui forme un contraste tranchant avec la liberté de manières des Français et des Allemands, me paraît fournir une preuve victorieuse en faveur de mon opinion.

"Le Basque est actif, persévérant et courageux, comme le sont toujours les hommes vigoureux et agiles; comme soldat, il n'est pas propre à servir dans la ligne, mais dans la petite guerre il se montre très actif, et devient redoutable à l'ennemi; son sang est chaud comme son climat, son courage inébranlable comme ses rochers, son attaque est impetueuse comme la mer qui baigne son pays. Le profond sentiment religieux qui l'anime ne suffit qu'à peine à tenir ses fougueuses passions en bride, et quelquefois la passion l'emporte sur la religion; cependant il ne connait pas les vengeances de l'Espagnol, et il est hospitalier comme lui. La profonde vénération avec laquelle les Basques parlent des morts, est un trait caractéristique de la nation. C'est probablement à cette vénération que tiennent les marques d'une douleur outrée qu'ils donnaient autrefois lors du décès d'un parent; ils s'arrachaient les cheveux, ils se flagellaient; le gouvernement a défendu ces excès sous des peines sévères. Le Basque aime avec passion la petite guerre, et les jeux où il peut déployer sa force et son adresse; il ne connaît d'autre patrie que ses montagnes, et aujourd'hui même il ne parle de la France que comme d'un pays etranger. * * Le Basque est probe dans le commerce; il ne montre point d'avidite, et se contente d'un gain modéré; il a échappé dans sa solitude aux vices dont les peuples limitrophes sont d'ordinaire entachés. Il est fort rare que le pâtre des contrées les plus élevées descende de ses montagnes, et si cela arrive ce n'est que pour aller vendre une chèvre *à la ville.* Ainsi, étranger à la culture intellectuelle et aux mœurs de notre siècle, il reste tout près de l'état primitif de nature, et il vit content dans son ignorance. Le cultivateur, plus aisé, fréquente les foires, et là il apprend un peu de Français; il n'en rapporte point dans ses vallées les nouvelles mœurs et la politesse de ses voisins, mais il reçoit l'étranger qui vient le visiter dans sa demeure avec l'hospitalité et la franche bonhomie des anciens temps."*

2. The ancient Lybians possessed the whole northern coast of Africa, from the confines of Egypt to the Straits and all the country thence reaching to the southward as far as it was known to the Greeks and Romans. It would appear that they were the only inhabitants of all these coasts before the age which preceded the foundation of Phœnician colonies among them; and the Carthaginians were called "Tyrii bilingues," because they spoke the Lybian as well as the Phœnician language; that is, the Berber and the Hebrew. The latter, however, became widely extended, and, as Gesenius has lately shown, was the idiom of inscriptions throughout Northern Africa.

* We may safely consider this éloge as rather poetical than exact.—ED.

The Lybian speech is still preserved among the rustic tribes who inhabit Mount Atlas, and in various parts of the interior. In the northern parts of Atlas, these people are called Berbers; in the southern tracts, they are the Shuluh, or Shilhas. In the hilly country belonging to Tunis, the Kabyles in Mount Auress speak the Showiah, which is another dialect of the same language.* Grammars and dictionaries of one dialect have been printed by the French Government, and we shall see that F. W. Newman, Esq. has successfully investigated the characters of the language generally, showing that it contains a peculiar and distinct vocabulary, with the addition of so great a number of Syro-Arabian words and grammatical forms, as to render the organisation of the language and its whole system of inflection entirely Syro-Arabian or Semitic.

One fact, not unimportant in its bearing on the early history of mankind, appears to have been rendered manifest by late researches in northern Africa: it is, a much wider extension over those regions, than was heretofore supposed to exist, of tribes bearing an unquestionable though sometimes remote affinity in language, and therefore probably in origin, to the Syro-Arabian or Shemite race. This denotes the very ancient dispersion of an Asiatic population over a great and central part of the African continent. I refer not, at present, to tribes of Arabian origin, or to such as can be supposed to have entered Africa subsequently to the era of Islam, but to races bearing indications of affinity to the Shemite stock, by far more ancient and more widely spread. The resemblances in language, and especially in the grammatical structure of languages to which I now allude, as existing between the African and the Shemite races, are approximations not to the modern but to the most ancient dialects of this latter family of nations. The idioms in which these phenomena display themselves may be termed with propriety Hebræo-African languages.

* The first specimen of this language ever obtained was a vocabulary of the Showiah given in Dr. Shaw's Travels.

### 1. *Berbers of the Northern Atlas.*

The mountains of Atlas are said to be inhabited by more than twenty different tribes, carrying on perpetual warfare against each other, tribe against tribe, and village against village. Hereditary feuds end only in the extermination of whole families. The tribes who live on the snowy mountains of Atlas dwell in caverns from November to April, and their exploits give origin to traditions and legends which terrify the people of the plains. They are very poor, and make plundering excursions in quest of the means of supporting life. They are a robust and active people.

1. The Berbers of the Higher Atlas are described by Lemprière, who calls them Brebes, as a very athletic, strong-featured people, patient, and accustomed to hardship and fatigue. He says that they seldom remove far from the spot of their abode; they shave the fore part of their heads, but suffer the hair to grow from the crown as far behind as the neck. Their only covering is a woollen garment without sleeves, fastened round the waist by a belt. These people, adds Lemprière, differ entirely from the Arabs and Moors, being the aboriginal inhabitants of the country, and in a great measure independent in their own mountain villages, where they feed cattle and hunt wild beasts.

2. The Shuluh, who are the mountaineers of the Northern Atlas, live in villages of houses made of stone and mud, with slate roofs, occasionally in tents, and even in caves. They are chiefly huntsmen, but cultivate the ground and rear bees. Leo Africanus reckons them as a part of the same race with the Berbers of the Northern Atlas; and according to M. Venture, their idiom, which they term Amazirgh, meaning the noble language, is a cognate dialect of the Berber speech. By Mr. Jackson it was considered as totally different; but evidence has been adduced by Captain Washington, in a memoir published in the Journals of the Royal Geographical Society, which

seems to prove that M. Venture's opinion was well-founded. The author has given a vocabulary collected by himself from the mouth of a native Shelha who had passed his life in Mount Atlas, which he has compared with the collections of Venture and others.

3. The Berbers of the Tunisian and Algerine territories are termed by the inhabitants of cities Kabyles, or Kabaily: they occupy all the hills which form the Lesser Atlas, the people of particular hills having the names of Beni-Sala, or Beni-Meissera, which mean "Children of Sala, or Meissera." They speak the Berber language, which is termed by them Showiah, and, in the interior of the country, are quite unacquainted with the Arabic. They live in huts made of the branches of trees, and covered with clay, which resemble the *magalia* of the old Numidians, spread in little groups over the sides of the mountains, and preserve the grain, the legumes, and other fruits which are the produce of their husbandry, in *matmoures*, or conical excavations, in the ground. They are the most industrious inhabitants of the Barbary States, and, besides tillage, work the mines contained in their mountains, and obtain lead, iron, and copper.

4. The Tuaryk are the people spread in various tribes, differing from each other in physical traits according to the climates of the countries where they dwell, through all the habitable parts of the great African plain of Sahara. They were accurately described by Leo Africanus, who had visited every part of their country; but their national affinity with the Berbers was not suspected till the time of Hornemann, to whom the discovery of this widely-extended race is to be attributed. The identification of the Tuaryk with the Berbers is due to Mr. Marsden. It has been proved by these writers that the Tuaryk nations reach eastward to the borders of Egypt. The Oasis of Ammon is inhabited by a people who speak their language.

The physical characters of the Berbers are described by M. Rozet. He says that the Berbers, or Kabyles of the Algerine territory, are of middle stature; their complexion is brown, and sometimes nearly black. "Les Berbères sont de taille moyenne; ils ont le teint brun et quelquefois noirâtre, les cheveux bruns et livres, rarement blonds: ils sont tous maigres, mais extrêmement robustes et nerveux; leur corps grêle est très bien-fait, et leur tournüre a une élégance que l'on ne trouve plus que dans les statues antiques. Ils ont la tête plus ronde que les Arabes, les traits du visage plus courts, mais aussi bien prononcés; ces beaux nez aquilins si communs chez ceux-ci sont rares chez les Berbères, l'expression de leur figure a quelque chose de sauvage et même de cruel; ils sont extrêmement actifs et fort intelligens."

The Shuluh in the mountains above Maroco are described by Captain Washington as lively, intelligent, well-formed, athletic men, not tall, without marked features, and with light complexions.

We must not omit the observation of Dr. Shaw respecting the Kabyles of the Tunisian country. He says that "the Kabyles, in general, are of a swarthy colour with dark hair; but those who inhabit the mountains of Auress, or Mons Aurarius, though they speak the same language, are of a fair and ruddy complexion, and their hair is of a deep yellow." Writers who labour under the prejudice which regards all physical characters as permanent, adopt the supposition, perfectly groundless as it is, that the xanthous Berbers of Mount Auress are the remains of the Vandals who were conquered by Belisarius. The Tuaryk are in some parts white, in others black, but without the features of Negroes.

The extension of this race through the Canary Islands is a curious and interesting discovery of modern times.

The Canary Islands and the neighbouring seas were explored by King Juba, of whose discoveries the younger Pliny has given us an account, as it appears, from Juba's own description: for this African prince was not only a navigator, but a celebrated writer on geography.

The first island, according to Juba, was named Ombrion; it had no vestiges of human habitation, but contained a mountain-lake: the second, and a small one adjoining, were termed Junonia; the next, called Capraria, abounded in lizards of great size. Nivaria, doubtless Teneriffe, was famed for perpetual snow and fogs; next to it was Canaria, so termed from its containing dogs of huge bulk, of which two were brought to Juba: here were found the remains of dwellings. All these islands abounded in fruits, and groves of palm-trees bearing dates, and filled with various birds and beasts.

It would appear from this account that the Canary Islands were but partially, if at all, inhabited in the time of Juba.

The modern history of the Canary Islands commences with their accidental discovery, in consequence of the shipwreck of a French vessel on the coast, between the years 1326 and 1334. Expeditions were afterwards made by the Spaniards for the sake of plunder and carrying off slaves; in one of which, the king and queen of Lancerote, and seventy of the inhabitants, were taken captive. At the beginning of the fifteenth century, a Norman baron, John de Bétancourt, subdued several of the islands, but Teneriffe was not brought under the yoke till ninety-five years afterwards. Here the native people, who were termed "Guanches," made a valiant resistance. The most instructive accounts of the Guanches are to be found in the narratives of some old voyagers who visited the Canary Islands during the time when they had been as yet but imperfectly conquered.

The population of Canaria Grande amounted to 9000, and that of Teneriffe to 5000 souls. The natives of the latter island are said to have been of great, and even gigantic stature. They were people of very simple habits and possessed of few arts; were ignorant of the use of metals, and are said to have ploughed the land by means of the horns of bullocks. They believed in a future state, and worshipped a Supreme Being, whom they termed Achahurahan,* the author and preserver of all good things. They also believed in a malignant being, termed Guayota, and placed the abode of the wicked in the burning crater of Teneriffe. They had a solemn institution of marriage, and various moral and social observances.

The practice of embalming bodies and laying them up in mummy-caves, or catacombs, in the sides of the mountains, is the most curious circumstance in the history of the Guanches; it is at least that which has attracted the greatest attention. The mummies were placed erect upon their feet against the sides of caves; chiefs had a staff placed in their hands, and a vessel of milk standing by them. Nicol, an English traveller, stated that he had seen 300 of these corpses together, of which he says that the flesh was dried up and the bodies as light as parchment. Scorey was assured that in the sepulchre of the kings of Guimar there was to be seen a skeleton measuring fifteen feet, the skull of which contained eighty teeth. Of late years we have obtained from Golberry, Blumenbach, and Humboldt, more correct accounts of these mummies, and of the mode employed in preparing them. The bodies were imbued with a sort of turpentine, and dried before a slow fire or in the sun. Their desiccation was so complete that the whole mummies were found to be remarkably light, and Blumenbach informs us that he possesses one which, with its integuments entire, weighs only seven and a half pounds, which is nearly one-third less than the weight of an entire skeleton of the same stature, recently stripped of the skin and muscular flesh. On opening these mummies, the remains of aromatic plants are discovered, among which the Chenopodium Ambrosioides is said to be con-

* According to a memoir by Sabin Berthelot, published in the "Mémoires de la Société Ethnologique," the name of the Supreme Being was Achaman at Teneriffe, and Acoran in Grand Canary. The name given in the text is an epithet meaning "the Great." See Mémoires, vol. i. part i. page 225.—ED.

stantly present. The corpses are decorated with small laces, on which are hung little dishes of baked earth.

M. Golberry has described a mummy in his possession, which he selected from among many others still remaining in his time in the mummy-caves in Teneriffe. Of this he says, "The hair was long and black, the skin dry and flexible, of a dark brown colour, the *back* and breast covered with hair, the belly and breast filled with a kind of grain resembling rice, the body wrapped in bandages of goat-skin."

Blumenbach thought he discovered some resemblance in the style of ornament between the mummies of the Guanches and those of the Egyptians. Strings of coral beads are found in both; but this may be an accidental resemblance, and the use of goat-skin instead of cloth, and the mode of filling the body and drying it, and all other particulars, differ essentially.

The incisores are worn down to truncated cones in the mummies of both nations. This may have arisen from their using similar food, or from both nations being in the practice of eating hard grains.

FIG. 70.

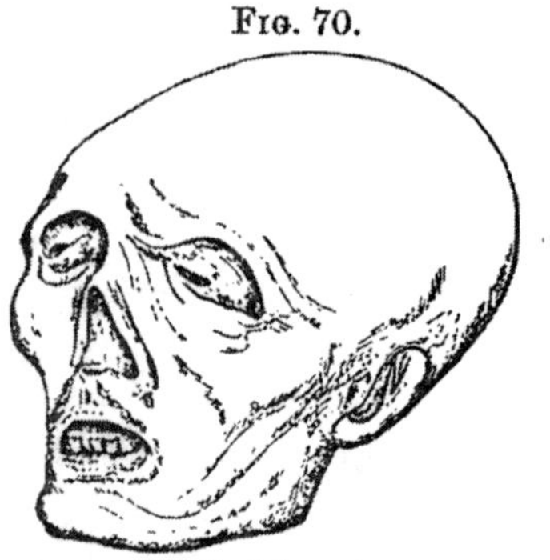

Skull of Guanche.

The sketch annexed, which is copied from Blumenbach's engraving, will afford a correct idea of the skulls of the Guanche mummies.

A few words accidentally preserved from the language of the exterminated tribe who formerly inhabited the Canary Islands, affords grounds for believing that they are of the Atlantic race.*

* It was maintained by M. Macedo, of Lisbon, in an ingenious memoir communicated to the Royal Geographical Society of London, that the idiom of the Guanches was a different language from that of the other islands and from the Berber dialects. The subject requires further elucidation.

The recent conquests of the French in Northern Africa have contributed to extend our acquaintance with the native people of that continent, and the successful investigations of M. d'Avezac have illustrated many obscure points in its geography. On African ethnology and the history of languages a new light has been thrown by the acute and penetrating researches of Mr. F. W. Newman, who has been the first to demonstrate, what many former writers have merely conjectured, that the language of the great and widely spread family of nations who extend over the whole north of Africa, from Mount Atlas, of which they are the original inhabitants, to the borders of Egypt and Abyssinia, is an ancient and a distinct branch of the Syro-Arabian or Semitic group, a coëval sister-language of the ancient Aramæan, the Hebrew or Canaanitish, and the primitive Arabian. The proofs on which this conclusion rests may be seen in several papers on the subject by Mr. F. W. Newman; but they are most fully developed in a memoir which that writer has contributed as an appendix to the fourth volume of my "Researches into the Physical History of Mankind."*

We have seen that there is good reason to be found in support of the opinion that this ancient race furnished the primitive population to the Canary Islands, as well as to the Balearic group, and several other islands in the Mediterranean.

The Tuaryk of the Great Desert are now well known to be of the Berber race. Their physical characters have been

* Appendix 2—"On the Structure of the Berber Language," by F. W. Newman, Esq.

The conclusions drawn by the author are, in his own words, as follows:— "On the whole, the evidence appears to show that the Berber is a Hebræo-African tongue, like the Ghyz and the Ambaric. With an enormous difference of vocabulary, its pervading genius is thoroughly the same, and, following grammatical peculiarities as our guide, the received doctrines on this subject would seem to justify the inference, that the Berbers are a race anciently connected by blood with the Canaanites and the Ethiopians" (p. 626).

already noticed. Great variations are observed among the different tribes of the Tuaryk, spread as they are through different climates. I shall add one observation on their physical characters.

The following account was communicated to me by Mr. Hodgson, the learned author of an excellent memoir on the Berber race, which was one of the first publications that drew the attention of the world to that interesting subject.* In the vast wilderness occupied by, or wandered over by the Berber Tuaryk, there are great variations of climate and situation. The natives of particular oases in the Great Desert are like the inhabitants of islands in the ocean. They never move in any considerable numbers from their native spot, nor are they visited by many strangers. They acquire, consequently, characteristics of physiognomy, through the agency of external conditions, the effect of which accumulates through many generations. In one of these oases—namely, that of Wadreag—Mr. Hodgson discovered that the people, though Berbers by the evidence of their language, which they speak with purity and correctness, were not only black, as many of the genuine Arabs of the country are known to be, but have features approaching those of Negroes, and hair like that which is characteristic of the Negro race. It was the opinion of Mr. Hodgson that these characteristics had been acquired, not as the result of the intermixture of races, which the local circumstances of the tribe seemed to him to preclude, but through the long-continued agency of physical causes upon a tribe of genuine Tuaryk origin, though the ordinary type of that race is almost similar to the Arabian.

Such a deviation in physical character is not wholly unparalleled, as the following remarks will tend to prove. I extract these from Dr. Wiseman's admirable lectures "On the Connexion between Science and Revealed Religion."

* W. B. Hodgson, Esq. on the Berber Language, "Transactions of the American Philosophical Society," 1834.

"The most curious example that I have met with of the spreading tendency to produce in one human race the characteristics of another, is mentioned by a recent traveller, almost the first who explored the Haüran, a district beyond the Jordan. He writes as follows :—'The family residing here—at Abu-el-Beady—in charge of the sanctuary, were remarkable for having, with the exception of the father only, Negro features—a deep black colour and crisped hair. My own opinion,' says Mr. Buckingham, was, that this must have been occasioned by their having been born of a Negress mother, as such persons are sometimes found among the Arabs in the relation of wives or concubines; but while I could entertain no doubt, from my own observation, that the present head of the family was a pure Arab of unmixed blood, I was also assured that both the males and females of the present and former generation were all pure Arabs by descent and marriage, and that a Negress had never been known, either as a wife or slave, in the history of the family. It is certainly a very marked peculiarity of the Arabs who inhabit the valley of the Jordan, that they have flatter features, darker skins, and coarser hair, than any other tribe—a peculiarity rather attributable, I conceive, to the constant and intense heat of that region, than to any other cause.' If all the facts and circumstances here given can be considered as sufficiently verified, we have certainly a very striking instance of approximation in individuals of one family to the distinctive characters of another, and of their being transmitted by descent."

[The late lamented traveller, Mr. James Richardson, passed in 1850 through much of the country inhabited by Tuaricks, who are frequently alluded to in the pages of his journal, which was published in the year 1853. In vol. i. p. 161, he says, "Tuarick is a generic title which scarcely implies even community of origin, assumed by nearly all the wandering people of the Sahara;" but the vocabularies which he collected, though slightly different in dialect, are all of the same language, which differs but little from Berber. Several entries in his journal contain descriptions of their personal characteristics, which assimilate them clearly to the people of north Africa. In p. 67 he says "they are finer and fairer than Fezzanees:" this was spoken of those seen by him in Fezzan; but much farther south (p. 242), he speaks of the agreeable countenance and tall stature of those he saw in Ahir. He says further of these, "some are of light olive complexion, with straight noses and thin lips, but others, indeed the greater number, approximate to the Negro features." This variety of physiognomy is mentioned by him several times (see i. 72, and ii. 153 and 157). An unexpected feature in the character of the Sahara population is mentioned at page 178 of Vol. i.: he is narrating his arrival at a spot where a quantity of dates for which he had paid were found packed up in the sand without any guard, and their place indicated by a piece of wood: he remarks, "had they been placed at the side of a well, and a hundred caravans had passed, no one would have touched them. It is a point of honour to steal nothing thus confided in the desert." The following note shows that the habit of fattening the women, so well known as prevailing on the coast, is also common among them:—"It is strange that some of the Tuarick women are enormously corpulent, whilst a corpulent woman is not found amongst the blacks" (i. 293).

One of the most interesting peculiarities about the Tuaricks, is the possession of a peculiar alphabet of unknown origin, which appears to have been in use for a very long period for engraving inscriptions upon rocks. Some specimens of such inscriptions were seen by Denham and Clapperton, and M. de Saulcy published and deciphered with some success one of considerable antiquity in the Journal Asiatique of 1843. The possession of this alphabet does not infer a literature, for there are no books, and hardly an attempt has been made to write it with a reed, pen, or other substitute: the only specimens the editor has seen were clumsy scratches hardly decipherable. A note from Richardson, sent to the Foreign Office in 1847, shows the practice. He says, "In the routes and highways of the Desert, as far as the Touarick authority extends, are found blocks of rock entirely covered with this character. I observed several of them . . . . . . In the towns which the Touaricks visit, we find many of their scribblings on the walls as well as their drawings. The house which I occupied at Ghadames had its walls covered with Touarick scribblings and drawings . . . . . . I have not seen any books in this character, and never heard that there were any to be found. The Touaricks seem principally to amuse themselves or pique one another with their native character, and do not use it for objects of utility: they are, however, very proud of it. The Touarick women one day asked me with exultation what writing it was, pointing to a great quantity of Touarick scribbling, and without waiting for a reply, exclaimed:—It is ours, it is better than yours, it is better than the books of the Arabs." The number of characters found may be about twenty; of these four or five are like those of the Himyaritic alphabet in shape and value. The others are unlike the characters of any alphabet the editor has seen.]

## CHAPTER XI.

### OF THE AFRICAN RACES INHABITING COUNTRIES BORDERING ON EGYPT.

HAVING described the ancient Egyptians in a preceding Chapter, I now proceed to give some brief account of numerous races in the Eastern parts of Africa, who, in their physical characters, bear some resemblance to that celebrated people. In some of these races, a certain approximation may be recognised to the type of the Negro.

The full development of all the peculiarities of organisation which are considered as characteristic of the Negro races are only to be found in the western districts of inter-tropical Africa. If we trace the intervening countries between Egypt and Senegambia, and carefully note the physical qualities of the inhabitants, we shall have no difficulty in recognising almost every degree or stage of deviation successively displayed, and showing a gradual transition from the characters of the Egyptian to those of the Negro, without any broadly marked line of abrupt separation. The characteristic type of one division of the human species here passes into another, and that by almost imperceptible degrees. This gradual change is not the result of the intermixtures of races on the confines of regions of old allotted to either separately. This might have been conjectured some years ago, and, in fact, it has often been said by those who sought the most obvious explanation of the phenomena. The intermediate tribes are not Mulattoes, or at all resembling Mulattoes: they have each their distinguishing features, which, besides their distinct languages, mark them out as races separate and peculiar, and not less distinct from the Negroes than white races themselves. These more accurate observations are the results of recent inquiries made on the spot by persons well skilled in natural history and comparative anatomy and physiology, and aware of the important bearing of such inquiries on the physical history of the human species. They were commenced by the scientific men who accompanied the army of the French republic in the Egyptian expedition. They have been followed up by later travellers. Some of the most extensive of these researches have been made by M. d'Abbadie, and I shall avail myself of the accurate discriminations made by this ingenious traveller.

In comparing the tribes of people among whom the intermediate gradation now to be described is discernible, I shall begin with some semi-barbarous races in the countries above Egypt, and reaching towards the borders of Abyssinia.*

*Of the African Nations bordering on Egypt and Abyssinia.*—The countries above Egypt are inhabited by two races of people resembling each other in physical characters, but of distinct language and origin. One is, perhaps, the aboriginal or native, the other a foreign tribe. I shall term them Eastern Nubians, or Nubians of the Red Sea, and Nubians of the Nile, or Berberines. All these tribes are people of a red-brown complexion, their colour in some instances approaching to black, but still different from the ebony hue of the Eastern Negroes. Their hair is often frizzled and thick, and is described to be even woolly; yet it is not precisely similar to the hair of the Negroes of Guinea.

1. The Eastern Nubians are tribes of roving people who inhabit the country between the Nile and the Red Sea: the northern division of this race are the Ababdeh, who reach northward in the eastern desert as far as Kosseir, and towards the parallel of Deir border on the Bishari. The Bishari reach thence towards the confines of Abyssinia. The mountain of Offa, fifteen days' journey distant from Assouan, is their chief seat. The Hadharebe are still farther southward, and reach to Souakin on the Red Sea. The Souakiny belong to this race. Macrizi speaks of these nations as partly Christians in his time: he calls them Bejawy, or Bejas. It appears that their country contained many churches, or religious establishments, previous to the devastation of Northern Africa by the apostles of Islam. The Bejas appear to have been the descendants of the people who in ancient times, under the name of Blemmyes, are described by Strabo and other writers as a powerful nation in the Nilotic countries. Being trouble-

* M. d'Abbadie's observations, made during his first journey to Abyssinia, were embodied in a memoir presented to the Institute. He had the kindness to communicate them to me in a letter written from Alexandria, and I shall cite them from his own hand.

some neighbours to the Roman governors of Egypt, they were driven out by Diocletian, who brought the Nobatæ of Libya to occupy their country. The latter are in all probability the Barabra, the present inhabitants of the valley of the Nile.

The present Bishari are extremely savage and inhospitable: they are said to drink the warm blood of living animals: they are for the most part nomadic, and live on flesh and milk.

The physical characters of this race have been described by many travellers who have visited some of their tribes. Among these are MM. Salt, Burckhardt, Du Bois-Aymé, Belzoni, and Wilkinson. The notices left by various writers as to their history have been carefully collected by M. Quatremère and the learned Professor Ritter. By these writers they are described as a handsome people, with beautiful features, fine expressive eyes, of slender and elegant form: their complexion is said to be a dark brown, or a dark chocolate colour. Belzoni, in describing the Ababdeh, says that "their hair is very crisp. Their head-dresses," he adds, "are very curious. Some are proud of having their hair long enough to reach below their ears, and then formed into curls, which are so entangled and matted with grease that they cannot be combed. That they may not derange their *coïffure*, they wear a piece of wood resembling a packing-needle, with which they scratch their heads." The figure of a Souakiny on the next page will serve as a specimen of their portrait.

M. d'Abbadie has given us the most accurate description of the Ababdeh. He says that the tribes near Ckossayr (Kosseir) have hair crisp (*crêpus*), but seven or eight centimeters long. He observes that the hair of these people is dressed in a peculiar way, which gives them a strange appearance. Their lips are not thick: "Leur nez est un peu gros dans le bas, et se rapproche du type Cophthe: leur teint est presque noir."

FIG. 71.

Souakiny Chief.

2. The Nubians of the Nile, or Barábra, or Berberines, are a people well known in Egypt, whither they resort as labourers from the higher country of the Nile. They inhabit the valley of that name from the southern limit of Egypt to Sennaar. They call themselves Barábra: by the Arabs they are named Núba. They are a people distinct from the Arabs and all the surrounding nations. They live on the banks of the Nile; and wherever there is any soil, they plant date-trees, set up wheels for irrigation, and sow dhourra and some leguminous plants. At Cairo, whither many of this race resort, they are prized for their honesty.

The Berberinés appear, from the most careful researches that have been made into their history, to be the descendants of the Nobatæ, who were brought fifteen centuries ago from an oasis in the western country, by Diocletian, to inhabit the valley of the Nile. They were converted to Christianity, such as it existed in those parts, but now profess Islam.

Blumenbach was strongly impressed by the resemblance between the Berberines and the pictures of the old Egyptians. The former are one of those races whose complexion is a mixture of red and black, and whose physical characters bear some analogy to those of the Egyptians; they are, however, much darker in colour than were that nation, though the shade of both varied. Brown, a most accurate writer, describes the people in the Island of Elephantine as black, but, in the opposite Assouan, of a red colour, with the features of Nubians, or Barábras. In a memoir on this race in the "Déscription de l'Egypte," they are said to be of a deep mahogany colour: "Les Barâbras se prévalent de cette nuance pour se ranger parmi les blancs. * * * Leur peau est d'un tissu extrêmement fin: sa couleur ne produit pas un effet désagréable; la nuance rouge qui y est mêlée leur donne un air de santé et de vie." Their hair is a very remarkable peculiarity of this race. "Ils diffèrent des Nègres," says the writer last cited, "par leurs cheveux, qui sont longs et légèrement crêpus, sans être laineux." In some of the children it is a mixture "de touffes noirs et de touffes blonds."* The blonde colour is not like that of Europeans, but resembles the hue of hair reddened by fire. Dr. Rüppell, who has given the most elaborate account of this people, among whom he made a long residence, says that "an attentive observation will enable us to recognise among the Barábras the old national physiognomy which their forefathers have marked upon colossal statues and the bas-reliefs of temples and sepulchres." Dr. Rüppell alludes to the sculptures found on the Nile above Egypt, which, however they may resemble the features of the Barábra, were not formed by

* Lepsius, writing at Korusko, about 23° N. lat., says, speaking of the Barábra around him, "This is an intelligent and honest race, of peaceable though far from slavish disposition, of handsome stature, and with shining reddish-brown skin," (p. 129, English translation, Lond. 1852).—ED.

their ancestors, but by people of the ancient Egyptian race. He thus describes their physiognomy:—"A long oval countenance; a beautiful curved nose, somewhat rounded towards the tip; lips rather thick, but not protruding excessively; a retreating chin; scanty beard; lively eyes; strongly frizzled, but never woolly, hair; a remarkably beautiful figure, generally of middle size; and a bronze colour, are the characteristics of the genuine Dongolawi. These same traits of physiognomy are generally found among the Ababdeh, the Bishari, a part of the inhabitants of the province of Shendi, and partly also among the Abyssinians."

The most interesting fact connected with this race is, that they appear, if we may place reliance on historical evidence, to furnish an instance of the transition from the physical character of the Negro to one very similar to that of the ancient Egyptians.* The Barábra are divided into three sections by their dialects, which are those of the Núba, the Kenous, and the Dongolawi, all living in the Nile valley. Núba is not a name assumed by the people themselves, but it is given them by the Arabs. This is, moreover, a term by which the Arabs designate all the black people coming from the slave-countries to the south of Sennáar. It must be observed that these black people are a class of Negroes who show, in their original state, some approximation to the characters of the Nilotic race. They are thus described by Burckhardt:—The Nouba distinguish themselves from the Negroes, among other circumstances, by the softness of their skin, which is smooth, while the palm of the hand in the true Negro feels like wood. He says, "Their noses

* A great deal of information has been collected relative to the history and ethnography of the Barábra, or Nubians, by MM. Costaz, Burckhardt, Waddington, Rüppell, Seetzen, Ritter, and other writers. Of the whole of this evidence I have endeavoured to give an analysis in the second volume of my "Researches into the Physical History of Mankind," to which I must refer my readers, as it cannot, for want of room, be introduced into the present work.

are less flat than those of the Negroes, their lips are less thick, and their cheek-bones not so prominent. Their hair is generally similar to that of Europeans, but stronger, and always curled; sometimes it is woolly. Their colour is less dark than that of the Negro, and has a coppery tinge."

Other writers agree with Burckhardt in their account of the Negroes of Qamâmyl and Bertat, the native region of the Núba. M. Cailliaud says that they have generally woolly hair, but that it is sometimes only crisp and curled. The particular district whence the Barábra issued appears to be Kordofan; and the race of Koldagi Negroes, as they are termed by travellers, still preserve and speak, as it would appear, a dialect of the Barábra language. Their idiom, at least, has a great affinity with that of the Barábra, a considerable part of its vocabulary, as far as it is yet known, being common to the two nations. Kordofan is probably the Oasis whence the Nobatæ, or Núba, originated.

The Nubian race, from which the Berberines descended, settled on the Nile fifteen centuries ago, and soon became partially civilised. In this region they have undergone a physical change which must be attributed, as it would appear, to the influence of external agencies different from those of their native land, and to that of civilisation, since they do not intermarry with the Arabs: the Blemmyes, who inhabited their country before they entered it, were driven out to make room for the Nobatæ. It must be admitted that almost all conclusions on such evidence as ethnology, or the history of races, affords, are liable to error, since we can seldom, or never, be perfectly sure that all the conditions of the problem are fully given, and that no circumstances have interfered to influence the results beyond those which are known. But if we claim allowance for such sources of fallacy, and the degree of uncertainty which they necessarily occasion, we may consider the history of the Nubian tribes as furnishing an example of change brought about during the lapse of ages in the physical character of a human race. If the Berberines, or Nilotic Nubians, are descended from the Koldagi Núba, we might be at liberty to suppose the Egyptian race, which resembled the Berberine, to have been the offspring of a similar stock; and this supposition would coincide with many facts which point out the Egyptians as an old African nation. On the other hand, the converse of this supposition is equally probable; namely, that the Núba themselves may have sprung from a tribe resembling the Egyptians. We have seen that among the lower animals the original characters of the wild stock are recovered when a tribe long domesticated has been restored to its primitive state: the uniformity of colour, the darkness of its hue, and the original conformation of the head and limbs, have reappeared. It is, therefore, just as probable that a tribe becoming barbarised in the forests of Central Africa should deviate from an original type similar to the Berberine or the Egyptian to that of the Koldagi Nubians, as that the change should take place in a contrary direction. The Núba themselves may be an offset from the original stock which first peopled Egypt and Nubia.

*Of some other Tribes of the Kwolla and Samhar of Intermediate Type.*—At Djouddah, or Jiddah, M. d'Abbadie saw a number of merchants and sailors from the neighbourhood of Souakin, whose national resemblance was very striking. He thus describes their physical structure:—"Tête d'une moyenne grandeur; les lèvres épaisses, la supérieure presque pointue dans le milieu; le nez élevé à la racine, qui est étroite, ensuite abaissé un peu, puis renflé vers le milieu, et enfin déprimé vers le bout, qui est rond; les yeux enfoncés; la paupière inférieure en poche, mais très petite; la supérieure entièrement cachée sous le sourcil, quand elle est ouverte; les pommettes saillantes et peu éloignées du menton, qui est court et retroussé; les dents très belles par le soin qu'ils ont de les brosser plusieurs fois par jour. Leur front a un léger creux horizontal au dessus des sourcils, puis est très renflé dans la partie nommée *sagacité comparative* par Spurzheim; cet auteur aurait appellé le front des gens de Sawakim large et philosophique. Oreille petite, à lobe non détaché; les joues grasses en haut mais élevées autour de menton, qui est dégarni de chairs; bras longs; peau bisriée mais presque noire; cheveux *laineux* et portés comme chez les Ababdés, mais formans une perruque encore plus épaisse; sourcils rares; peau fine et ayant peu de poils; yeux bruns et enfoncés; cuisses moins grèles que chez beaucoup d'Arabes; point de mollet, la partie antérieure du tibia étant aussi saillante que le derrier de la jambe."

M. d'Abbadie describes another race, namely the inhabitants of Samhar and the Somali, who, as he says, have a form of body resembling the European, but are in complexion almost perfectly black, and have thick lips, and resemble in their hair the Ababdeh. "Leurs cheveux épais naturellement frisés se projettent derrière la tête en épaisse perruque comme la chaume d'un toit." Some of the Shoho have grey or blue eyes. "Comme chez les Nubiens, la peau des Chohou et des Habab est très douce, quoique presque noire." It seems from Burckhardt's observations, that this quality of the skin is considered as characteristic of these woolly-haired blacks termed Núba, and is held to distinguish them from genuine Negroes. But among those races who are considered as truly Negroes, we find some of the characters of the Ababdeh still prevalent. A proof will be found in the following description of the race of Negroes who inhabit the Kwolla, or belt of low country skirting Abyssinia towards the north. I cite M. d'Abbadie's words:—

"Ces Nègres forment l'un des races intermédiaires qui offrent la transition du type Européen à celui du noir de Guinée. J'ai dessiné ainsi leur portrait:—

"Oreille en arrière du plan passant par le milieu de la tête; lèvres épaisses; cheveux laineux, absolument comme chez les Chohou ou Habab; racine du nez sensiblement plus aplâti, mais beaucoup moins que dans le Nègre de l'occident; nez court et légèrement aquilin et s'approchant du camus; menton fuyant un peu en arrière; visage paraissant peu intelligent, mais bien au-dessus de celle des Nègres en général. Leur langue s'appelle Napat, et l'on dit qu'ils ont plusieurs grandes villes."*

## CHAPTER XII.

### ABYSSINS, OR RACES INHABITING THE HIGHLANDS OF ABYSSINIA.

THERE is no part of Africa, Egypt being excepted, the history of which is connected with so many objects of curiosity as Abyssinia. A region of Alpine mountains, ever difficult of access by its nature and peculiar situation, concealing in its bosom the long-sought sources of the Nile, and the still more mysterious origin of its singular people, Abyssinia has alone preserved, in the heart of Africa, and in the midst of Moslem and Pagan nations, its peculiar literature and its ancient Christian Church. What is still more remarkable, it has preserved extensive remains of a previously existing and widely spread Judaism, and, with a language approaching more than any living tongue to the pure Hebrew, a state of manners, and a peculiar character in its people, which represents in these later days the habits and customs of the ancient Israelites in the times of Gideon and of Joshua. So striking is the resemblance between the modern Abyssinians and the Hebrews of old,

* The interesting journals of Ferdinand Werne, a German in the service of the Pasha of Egypt, who was for a long period in the countries between Egypt and Abyssinia, add very much to our knowledge of the people inhabiting them, and something additional about them will be found in Parkyns's "Life in Abyssinia."—ED.

that we can hardly look upon them but as branches of one nation; and if we had not convincing evidence to the contrary, and knew not for certain that the Abramidæ originated in Chaldæa, and to the northward and eastward of Palestine, we might frame a very probable hypothesis which should bring them down as a band of wandering shepherds from the mountains of Habesh, and identify them with the Pastor kings, who, according to Manetho, multiplied their bands in the land of the Pharaohs, and being, after some centuries, expelled thence by the will of the gods, sought refuge in Judea, and built the walls of Jerusalem. Such an hypothesis would explain the existence of an almost Israelitish people, and the preservation of a language so nearly approaching to the Hebrew, in intertropical Africa. It is certainly untrue; but we find no other easy explanation of the facts which the history of Abyssinia presents, and particularly of the early extension of the Jewish religion and customs through that country; for the legend which makes the royal house of Menilek descend from Solomon and the Queen of Sheba is as idle a story as ever monks invented to abuse the reverent ignorance of their lay brethren.*

The highland of Abyssinia, compared by Humboldt to the lofty plain of Quito, advances many degrees to the northward of the great chain of mountains which traverses

* The entire absence of any Semitic language in Asia beyond the corner of that vast continent bordering on Africa, and the existence of several idioms in Africa having grammatical and glossarial affinities with the Semitic tongues, radical and obviously not borrowed, in addition to the Ghíz and widespread Arabic in the same continent, prove the so-called Semitic languages to be at least as much African as Asiatic. The ethnographical area which embraces the S.W. corner of Asia and the N.E. of Africa is decidedly one, and the geographical area may be considered in the same light, for the Red Sea rather unites than divides the nations on its borders. These facts will account for the similarity of idioms and manners alluded to in the text, without any hypothesis at all irreconcileable with the language and narrative of the Sacred writer. This opinion was maintained on somewhat different grounds by my friend Dr. Beke in his suggestive Origines Biblicæ, published in 1834.—ED.

Africa from east to west. Abyssinia, according to Tellez, is called by the inhabitants "Alberogran," or the Lofty Plain. By this epithet they contrast it with the Kwolla, or low country which everywhere surrounds it, except on the southern side. It is compared by them to the flower of the "Denguelet," which displays a magnificent corolla environed by thorns, in allusion to the many barbarous tribes who inhabit the surrounding valleys and low plains. Behind this country, the plain of Narea, or Enarea, reaches still farther south, and serves, as Ritter observes, to connect Habesh with the still more lofty mountains of Kaffa, and the great elevated region of Central Africa. The Highlands of Abyssinia, properly so termed, reach from the southern province of Shoa and Efat, which are not far distant from Enarea, under the ninth degree, to Tcherkin and Waldubba under the fourteenth, of northern latitude, where they make a sudden and precipitate descent into the low forests occupied by Shangalla Negroes. The greater part of the Upland of Habesh is a country of Alpine pasturage: it contains some cultivated plains, and scarcely any forests; but it is in certain seasons abundantly watered by numerous fertilising streams. It feeds innumerable herds of oxen and horses, and races of men, vigorous, handsome, active, and intelligent, ever more addicted to the arts of war than those of peace, and who, as Ludolph says, lay down their arms only when forced by the return of intertropical rains.*

Abyssinia was for ages united under one governor, the Negus, or emperor, who resided in the earliest periods at Axum, the ancient capital of Tugray or Tigré, and for some centuries past at Gondar, in the more central part of the country. It is divided into several provinces, or kingdoms, and inhabited by several races of people, who, though in physical and moral characteristics similar, are distinguished from each other by that great mark of original

---

* Ludolph, "Hist. Æthiop." lib. i. Ritter, "Erdkunde," th. i. c. 3.

diversity—an essential difference of speech. Untouched by the ancient culture of Egypt and Ethiopia, Abyssinia seems to have derived all its earliest improvement from the Shemite nations, on the opposite side of the Arabian Gulf. Thence came, probably, the syllabic characters, long, as it appears, common to the Abyssinians and their kinsmen, the Homerites of Southern Yemen. The coast of the Red Sea, and the lower tracts along its border, are the Samhar, or Samhara: thence the traveller ascends the lofty hills of Assauli and Taranta, into the western Upland of Tugray: that was the country where, in Axum the metropolis of the Negus, arts were first cultivated; and there, although Judaism, as it seems, prevailed far and wide, the knowledge of Grecian sculpture and Greek letters, and with it the polytheistic mythology of Egyptian Greeks, had penetrated during the age of the Ptolemies, while foreign merchants frequently visited Adulé, and other ports in the Red Sea. But the old Ghíz, or Hebræo-Ethiopic, continued to be the language of the people of Axum down to the time and long after the arrival of Frumentius, who was consecrated by the great Athanasius, and became the apostle of Abyssinia. Frumentius translated the Sacred Scriptures into the Ghíz, or old dialect spoken among the Tugrayans in the eastern provinces, who were then the dominant tribe; while the Falasha, in the interior parts, retained Judaism; and the southern tribes, the Agows and others, adhered to their original African Paganism and the adoration of the Nile. The Tugrayans, to the eastward of the Astaboras, or Takazay, were the genuine Shemite, or perhaps Cushite, Abyssinians. The Amharas, a race who spoke the Amharic, inhabited the most extensive province of Abyssinia; and in their country is Gondar, which became the seat of power in a later age. The Amharic contains a great mixture of Arabic and of Ghíz; but the most learned philologists who have studied it regard it as an idiom not fundamentally Syro-Arabian. This point is not fully decided, and on its decision will depend that of the inquiry whether the Amharas were a Shemite people, or a genuine African race. In the present state of our knowledge, this last opinion is the most probable; and it may be extended to all the other nations who, together with the Tugrayans, constituted the subjects of the Negus.*

Abyssinia, as it is well known, has been overrun, and its southern parts conquered, in later times, by the Galla, a barbarous people who surround it on its southern, south-western, and south-eastern sides, and who now form a great part of its inhabitants, and seem likely ere long to exterminate or to swallow up the other races in their own greater numbers and increasing population.

Besides the two principal languages, which are, as we have said, the Ghiz and the Amharic, the following idioms, several of them, as far as we can judge from the specimens given of them by Mr. Salt, constituting not merely dialects, but entirely distinct languages,† which must be looked upon as marking so many distinct races, are now spoken within the limits of Abyssinia:—

---

* I have no doubt that the Amharic is fundamentally an idiom of the class called in the text Syro-Arabian. The Shoa dialect has, it is true, many Galla words in it; but those words, in the judgment of Dr. Krapf, whose long residence in Shoa, where he became familiar with the Amharic language, and subsequent acquaintance with the Galla's, south of the Equator, give him a right to pronounce, have been adopted from the Gallas. The reasons for this judgment are that many Gallas inhabit Shoa, while no Amharas ever go to the Galla country; that the common words in question are found in remote provinces where Galla only is spoken; and that they are not known in the more distant provinces of Abyssinia where the best Amharic is spoken. See Preface to East African Vocabularies, p. ix. But even a grammatical connection with the Galla tongue would not impugn the notion of the Syro-Arabian character of the Amharic language. The opinion of Mr. Newman, mentioned in p. 319 of this work, that the Galla is allied to the Berber, a Syro-Arabian language, is confirmed by the subsequent publication of Tutschek's Galla Grammar.—ED.

† The more extended vocabularies published by Dr. Beke do not confirm this radical distinction; but, until we have grammars of these languages, the question must be left in doubt. At present, the evidence seems in favour of connection rather than separation.—ED.

1. The language of the Agows in the province of Avergale. 2. That of the Agows, to the westward of Matscha, which is different from the former. 3 and 4. The idioms spoken by the Falasha and Waitos, who inhabit the provinces of Janfangera and Fangia.

Besides these, two other languages are enumerated by Rüppell: that of the Shoho, called by Salt and others Shiho and Hazorta, who are shepherd tribes on the north-western boundaries of Abyssinia, and on the borders of the Samhar;* secondly, the dialect of the people called Shangalla of the Takazay.†

[Some interesting notices of the Shohos or Sahos will be found in Mr. Parkyns's Notes collected during a three years' residence in Abyssinia. They are Mahometans, and wholly pastoral, "refusing to have any hand in the cultivation of the soil." Mr. Parkyns visited some of their camps, or rather villages, for the Sahos construct dwellings rather more solid than pastoral tribes usually require, though not much more stationary: he says, "their huts are formed of straw and branches of trees, neatly enough fashioned and thatched; they are placed so as to form a circle, in which the cattle are penned for the night; one or two spaces are left as entrances, and these are closed at night by bushes strewed before them." There is a sort of division of labour between them and the Abyssinians, which Mr. Parkyns notices as highly advantageous to both parties.

* The Shoho inhabit the foot of the Assauli and Taranta Mountains, and some districts farther southward. Rüppell conjectures that they were originally a tribe of Galla, since they resemble that people in physical characters. They have, however, a distinct language, of which Mr. Salt has given a specimen, *vide* p. 296, *infrà*. The Hazorta are a subdivision of the same race. Salt calls them Shiho. According to Rüppell, the name of the Hazorta is properly Za-horta.—See Rüppell, "Reise in Abyssinien," b. i. s. 263.

† The latter are described as Negroes by Bruce and Salt; but Rüppell says that in physical characters they resemble the Shoho. The fact that they bear the designation in Abyssinia of Shangalla of the Takazay, indicates that they are considered in that country as a kind of Negroes, since the name of Shangalla is equivalent to that of Negro.

"The Abyssinians being entirely agricultural, the rich owners of oxen among them entrust their animals, after their services at the plough are no longer required, to the care of a Shoho, who pastures them for the remainder of the year, receiving in payment a quantity of corn on their safe return. On the other hand, rich Shohos, owners of vast herds of cattle, lend out their oxen to poor Christians, who cannot afford to purchase any for themselves. The Abyssinian, owner of the land, has the entire labour and management of the crop, while the Shoho, owner of the oxen, has a share of the harvest," (Vol. i. p. 126.) Their costume differs from that of the Abyssinians, as may be anticipated from the difference of habits; but we should scarcely be prepared for the exhibition of vanity mentioned in the "Notes." "The tressed hair of the Abyssinian Christian contrasts strongly with the bushy wig of the Shoho, who arranges his woolly hair into two large tufts, one of which is on the top of the head, and the other behind. By way of ornament a pin or scratcher is stuck through the front tuft . . . . . . It is amusing to see with what a careful air of self-satisfaction a young Shoho will draw out his long hair-pin, and, after having passed it two or three times through his hair, replace it in the fore bush immediately over his forehead, with as much of it protruding as he can possibly manage without its falling, at the same time smiling most contentedly at nothing at all, or giving vent to a shrill whistle, as if driving his cattle,—perhaps to let the world know that he is the owner of a good herd. The Abyssinians wear breeches and large belts, instead of which the Shohos sometimes substitute a kilt of cotton stuff, which falls a little below the knee, or content themselves with the "tobe," or cloth, alone, which in this case is made to answer the double purpose of coat and trowsers. Being passed first round the body, so as to cover the lower extremities, the ends are crossed on the breast and thrown over the shoulders. For convenience it is occasionally tied at the back of the neck" (Vol. i. p. 128.)

The Saho language is shown by Ewald to be Semitic; see Zeitschrift für die Kunde des Morgenlandes, vol. v. p. 410, and Journal Asiatique, 1843, pt. 2, p. 108. But it is Semitic in the same way as Berber, Tuaric, and Amharic, and not as Arabic and Ghíz. We shall perhaps find families of Semitic languages allied to each other in the same degree as are the Slavic, Germanic, and Latin families of the great Indo-Germanic class.—This was surmised by the Author, as is seen in the preceding pages.]

### *Physical Characters of the Abyssinians.*

The Abyssinians are reckoned among black races. Arabian historians who narrate the wars between the old princes of Yemen and the Negus, term them *Blacks*, and apply to them epithets which Schultens translated "Æthiopes crispâ tortilique comâ." One of the Arabian princes, suing to the Persian King, entreats him to drive out these crows, who are hateful to his countrymen. Burckhardt says that the female Abyssinians are the most beautiful of all *black* women.

Dr. Rüppell informs us that there are two physical types prevalent among the Abyssinians, the Galla and the Shangalla being excluded from that designation. The greater number, he says, are a finely formed people of the European type, having a countenance and features precisely resembling those of the Bedouins of Arabia. The characteristic of their exterior consists principally in an oval shape of the face; a finely-pointed nose; a well-proportioned mouth, with lips of moderate thickness, not in the least turned out; lively eyes; well-placed teeth; somewhat curled or smooth hair; and a middle stature. The greater number of the inhabitants of the high mountains of Samen, and of the plains around Lake Tzana, as well as the Falasha or Jews, the heathen Gamaut, and the Agows, notwithstanding the variety of their dialects, belong to this class. A second numerous division of the Abyssinian people, according to the same traveller, is identified, so far as physical traits are concerned, with the race which he has designated under the name of Ethiopian. "This last type," says Dr. Rüppell, "is distinguished chiefly by a less acute and uniformly somewhat flattened nose; by thick lips; by long and not very sparkling eyes; and by very strongly crisped and almost woolly hair, which stands very thickly upon the head. A part of the inhabitants of the Abyssinian coast, of the province Hamasen, and other districts along the northern borders of Abyssinia, belongs to this Ethiopian race." These are the characters which Rüppell, in a previous work, had ascribed to the Berberines of the Nile and the Ababdeh. He says that the portrait of the Suakiny Arab, given in Lord Valentia's travels, from which the figure in page 283 is taken, is a very good exemplification of this type of countenance and general character.

This last physical type, which Rüppell terms Ethiopian, and declares to be common to a considerable part of the Abyssinian and Nubian tribes, with the Berberines, and Ababdeh, and Bishari, is precisely that character of physiognomy which, by Larrey and most other writers, is described as the prevalent type of the Abyssinian countenance. Baron Larrey, in particular, who has entered very fully into the physical history of these races, describes one type as common to the Copts, or native Egyptian race, the Barábra, or Berberines, and the Abyssinians; and he separates this by a broad line from the character peculiar to the Negro races, and by almost as broad a line from that of the Arabian. I shall cite his observations, as he may be considered to be the greatest authority on this subject. The Egyptians, or Copts, who, as he says, form one branch of this assemblage of races, have a "yellow, dusky complexion, like that of the Abyssinians. Their countenance is full, without being puffed; their eyes are beautiful, clear, almond-shaped, languishing; their cheek-bones are projecting; their noses nearly straight, rounded at the point; their nostrils dilated; mouth of moderate size; their lips thick; their teeth white, regular, and scarcely projecting; their beard and hair black and crisp." In all these characters, the Egyptians, according to Larrey, agree with the Abyssinians, and are distinguished from the Negroes. "En effet les Nègres Africains ont les dents plus larges, plus avancées, les arcades alvéolaires plus étendues et plus prononcées, les lèvres plus épaisses, renversées, et la bouche plus fendue: ils ont aussi les pommettes moins saillantes, les joues plus petites, et les yeux plus ternes et plus ronds, et leurs cheveux sont lanugineux." With this description he contrasts that of the Abyssinians, who are distinguished by large eyes and a fine expression of countenance, the inner corner of the eye displaying a slight curve; the cheek-bones are more prominent, and form, with the marked and acute angle of the jaw and the corner of the mouth, a more regular triangle; the lips are thick, without being turned out, as in the Negroes; and the teeth are well formed, regular, and less projecting; the alveolar edges are less extensive. The complexion of Abyssinians is the colour of copper. "These characters," says M. Larrey, "are common, with slight shades of difference, to the Abyssinians and the Copts. They are likewise recognised in the statues of the ancient Egyptians, and, above all, in the Sphinx, as well as several of the Egyptian mummies." "Pour vérifier ces faits," he continues, "j'ai recueilli un certain nombre de crânes dans plusieurs cimetières des Qobtes dont la démolition avoit été nécessitée par les travaux publics. Je les ai comparés avec ceux des autres races, surtout avec ceux de quelques Abyssins et Ethiopiens, et je me suis convaincu que ces deux espèces de crânes présentent à peu près les mêmes formes." He says that the mummy-heads found at Saqqarah displayed precisely the same character, namely the prominence of the cheek-bones and of the zygomatic arches, the peculiar shape of the nasal

fossæ, and the relatively slight projection in the alveolar edges, when compared with the corresponding structure in the Negro skull.

The figures in the coloured plate which accompanies this page were both taken under the inspection of M. d'Abbadie, and they afford characteristic specimens of the two principal varieties of type among the Abyssinian races. The left hand figure represents the countenance which is termed by Rüppell Ethiopian, and which displays, as may be seen, a notable approach to the character of the Negro, or at least a decided deviation from the prevailing physiognomy of the European and the Arabian; the other is the portrait of a native of Shoa, and it exhibits a countenance which, though peculiar, has nothing decidedly African: its type is that of the Shemite race of Abyssinians. The well-known portrait of the learned Abbas Gregorius seems to belong to the former class of physiognomical forms. Gregorius was a man of genuine Amharic descent; his broad, somewhat flattened features resembled those of a Copt: he had a complexion almost black, and, like the rest of his countrymen, as Ludolph says, crisp or frizzled hair. The portrait in the plate displays tolerably well these peculiarities.

The most remarkable additions that have been made, of late years, to our acquaintance with several branches of the human family, are connected with the history of races inhabiting the north-eastern parts of Africa. The geography of that region of the world, and particularly of Abyssinia and the countries lying to the southward of the Abyssinian empire, has lately been explored by missionaries and other travellers, and more precise and satisfactory accounts have been given of the physical and moral characteristics, and likewise of the languages of several widely-spread races, than had been previously obtained by European geographers or students of philology.

Most of the recent observations as to the elevation, site, and local position of the high country of which Abyssinia is a part, are found to elucidate and coincide with the general idea formed of the nature of that region by Professor Ritter. It was supposed by that able and learned writer, that a lofty tract, consisting partly of separate mountain-chains and in part of vast uplands or high level plains, projects in the eastern part of the continent from the northern border of the great central African plateau, or stands as it were before it, like an advancing buttress, reaching out into the region of low countries towards the north, nearly as the elevated terrass of Tibet and Bhútan advances southward in front of the great plateau of High Central Asia.* In this region, elevated in its general surface, several insulated mountains attain a still greater height. The peak of Samen, called Amba-Hai, is supposed to rise 16,000 feet above the level of the sea; and though only 13 degrees distant from the equator, it is covered with a mantle of perpetual snow. The mountains of Lasta, Angot, and the hilly tracts of the southern Shoa, where, though far within the tropics, hail, frost, and snow are often seen, cannot, as it was observed by one of the most intelligent writers on the geography of Africa, be of less elevation than that of 12,000 feet. The sources of the Abyssinian Nile were supposed by Bruce to be 10,340 feet above the sea; Mount Amid rises to a much greater height. In Kaffa, though much nearer to the equinoctial line, there are mountains covered with eternal snow. Such, according to Ptolemy, were the Mountains of the Moon, which, almost under the equator, are supposed to surround the sources of the Bahr-el-Abiad, or the Greater Nile.†

From these regions of high mountains the great rivers of Eastern Africa take their rise, and flow in all directions

* Erdkunde der Erde, von Karl Ritter, Theil i. Erdkunde von Afrika.

† See the learned "Geographical Memoir on Eastern and Central Africa," prefixed to the published journals of the Rev. MM. Krapf and Isenberg, by James Macqueen, Esq.; and a recent memoir, by the same distinguished geographer, in "Blackwood's Magazine," published June 1844.

into the lower circumjacent tracts, excepting only at the south-western angle, where a continuous chain of elevated lands is supposed to connect the Abyssinian plateau with the more extensive highland regions of Central Africa. Towards the north, rivers of considerable magnitude, the Mareb, the Tacazay or Astaboras, and several contributary streams of the Bahr-el-Azrek, or Blue River, carry their waters to the common channel of the Nile. Further westward the Bahr-Seboth, or Red River, and several inferior streams descend, and contribute to fill the channel of the Bahr-el-Abiad. From the same high country the Hawāsh flows down due east, and, passing through the low plains of Aussa, falls into the gulf of Tajura, near the straits of Bab-el-Mandeb. To the southward, beyond the high mountains of Enarea and Kaffa (one said to be the native country of the tea-plant, the other of the coffee-tree), flow several large rivers, the contributary streams of the Gochob, the Juba of the ancients. The Juba, one of the principal rivers of Africa, winds its long course through the plains of the Galla and the Somáli, to which nations it serves as a dividing boundary, and empties itself into the Indian Ocean. The Quilimance and the Melinda, and other rivers of the eastern African coast flowing towards the south-east, derive, as it is supposed, their waters from the neighbouring highlands, and pour them into the sea which washes the coast of Mosambique.

In these regions, and in the countries extending further towards the south and west, are vast plains reaching through many hundreds of miles; in other parts mountain-chains, concealing secluded valleys, where umbrageous forests mitigate the heat of an intertropical climate. The former are the pasture-fields of nomadic and equestrian hordes, warlike and predatory wanderers, who, like the Scythians or Turanians of High Asia, make formidable incursions on the settled inhabitants of the surrounding countries. The latter conceal, in inaccessible fastnesses, various insulated human families, as yet little known to the rest of mankind, some of which, from their habits and their singular shape of body and diminutive stature, are supposed to be the types of the gold-collecting gryphons, and of other celebrated monsters of ancient fable. A few more years will probably dispel all the mists of obscurity which yet impend over the almost unknown wildernesses of remote Africa, and will clear up all that is mysterious and portentous in that region of the world. It will then probably appear that nothing is there really existing that contradicts the results of observations made in other countries.*

## CHAPTER XIII.

### OF THE WOOLLY-HAIRED RACES OF EASTERN AFRICA.

*The Shangalla—the Doko.*

In countries so different from each other in the conditions of climate and local situation, and in all the agencies which are known to exercise an influence on the physical development of human races, it would be contrary to all expectation if there were not found to be great varieties in the form and complexion, and in the whole bodily and mental constitution, of the inhabitants. Such varieties have, in fact, been discovered, and they reach to a great extent. Many of the tribes who inhabit elevated countries are people of fine form, noble stature, and great personal beauty. In the highlands approaching Enarea the people are of light colour; according to the late observations of

* The travels of Dr. Beke, Mansfield Parkyns, and Dr. Krapf, have done much to fulfil the expectations of the author; and so far as the Abyssinians proper are concerned, there is all the information required for a good monograph: but though much has been done, the time is not yet come for a decision upon the tribes in the vicinity. I am of opinion that they will be reduced to two great families,—that of the predatory Gallas, and that of the more settled and industrious Kafirs.—Ed.

M. d'Abbadie, many tribes are of a light copper-colour, and some very fair. Bruce long ago declared that the people of these elevated tracts are not darker in complexion than the natives of the southern countries of Europe: he alludes to the natives of Kaffa and Enarea, who speak a dialect of the Galla language. The description given by the last-mentioned writer of the Shangalla nations is strikingly graphical and interesting. The name of Shangalla belongs to the indigenous hordes who inhabit the Kwolla, or the deep woody valleys which surround on almost every side the highlands of Abyssinia. They are Negroes of a jet-black colour, and strongly-characterised features. These tribes were first described by Ptolemy, who classified them by the varieties of their sustenance, and called them Rhizophagi, Elephantophagi, Acridophagi, and Struthiophagi, or people who live on roots, or on the flesh of elephants, upon locusts, and ostriches. Others, as Bruce says, feed on a beautiful species of lizard. "During the fair half of the year," says Mr. Bruce, "when the Shangalla live under the shade of trees, they bend the branches downwards and cover them with the skins of beasts. Every tree is then a house, under which dwell a multitude of black inhabitants till the tropical rains begin. It is then they hunt the elephant, which they kill by various devices, as well as the rhinoceros and other large creatures. Where water and river-horses abound, they kill them with the same industry. Where the trees are thickest and the water in largest pools, there the most populous nations live, who have often defeated the royal armies of Abyssinia."

[Mr. Parkyns, during a residence at Rohabaita (lat. 14¾° N.) had an opportunity of seeing a very little, but hearing a good deal of the Shangallas, whom he calls more usually Bareas, but considering the terms identical.* His attempt

* "Life in Abyssinia," vol. i. p. 336. Mr. Parkyns informs us that their real name is Bàza, and that the word Barea means "slave" in the Abyssinian languages.—ED.

at penetrating into their country by the aid of the Bideles, a people of like habits living east of the Barea, was frustrated by a war or rather "razzia"* of the Abyssinians. He learned something, however, from a Barea prisoner taken in the war, and something from an Abyssinian who had been captured and had made his escape. The tribes described reside between the Tacazay and the Mareb. They dress very scantily, their only covering being a mere cloth of one-third of the size of the Abyssinian "quarry," itself a scanty covering; they wear also sandals, which are not used by the Abyssinians. They are armed with a small round shield, having a boss for the hand, a double-edged sword, and a spear; their swords are made in Europe, and procured by the Shangallas from the Egyptian caravans, which diverge into the various regions of Africa. They are said by the Abyssinians to be Pagans, but from their use of Mahometan names and practice of circumcision, Mr. Parkyns is of opinion that they are half-converted Moslim. They cultivate dokhon for food, but eat also of many animals forbidden by Mahometans, and considered unclean even by Abyssinians. They are very brave, strong, active, and hardy, and are said to be honest and trustworthy. The anecdotes told of these tribes, and of their conduct in the incursions mutually made by the two parties, though related by their deadly foes, all shew them in a favourable light, in so far as regards boldness, activity, and skill in savage warfare, and leave an impression of a higher sense of honour and less degree of barbarity than would be expected from their condition and generally received character.]

To this class of nations, the Shangalla, must be referred the tribe of Negroes said to inhabit forest-lands to

* The French occupation of Algiers has almost naturalised this word. It should properly be ghazia, but the guttural Arabic consonant is so much like a very common though vulgar pronunciation of the French r (the Northumberland burr) that the colonists have constantly so rendered it.
ED.

the southward of Enarea, in the vicinity of the river Gochob, or the Juba of the ancients. These are the Dokos, or Pygmies, of whom accounts have appeared in several publications—accounts which are so singular, that by some they have been regarded as fabulous. The first notice of this people was given by the missionary Dr. Krapf: it was inserted in a geographical memoir on the river Gochob and the countries of Kaffa and Enarea, which was presented to the Egyptian Society at Cairo. This was thought by Professor Ritter a very important addition to African geography, and it was translated by him, and inserted in the "Monathsbericht" of the Geographical Society at Berlin, whence the particulars of this relation found their way into various English periodical papers. The account of the Dokos cannot be perused without exciting a strong suspicion in the mind of the reader that parts of the story have been exaggerated or distorted. This, however, is not a sufficient ground for rejecting the whole as a mere fiction, which some have done. In order that the reader may have an opportunity of exercising his own judgment, I shall abstract the most important part of the narrative, without selecting what may appear to myself most credible and rejecting the remainder.

It must be observed that the whole story rests on the credit of a Galla slave named Dilbo, a native of Enarea, who had personally visited the country to the south-west of Kaffa, where the Dokos are said to have their dwelling.*

* An account of Dilbo has been given by Dr. Beke in the twelfth volume of the "Journal of the Royal Geographical Society." He is considered by that intelligent traveller as a person worthy of credit.—Dr. Prichard.

In a paper "on the Geographical Distribution of the Languages of Abessinia," printed at Edinburgh in 1849, Dr. Beke, remarking upon this wonderful account, says: "When [Dilbo] was questioned by Dr. Krapf and myself, his statements were such as to entitle him to full credit; and I am afraid that he was subsequently induced to enter into these fanciful and exaggerated details, by a feeling not uncommon among uneducated persons, when *pressed* to furnish information, that the more wonderful they make their story, the greater praise they will obtain,—and, probably, also the

From evidence given by the same African, a great number of particulars have been taken relative to the geography of the neighbouring places, and of the course of the river Gochob. These geographical notices bear every indication of accuracy, and they have been received without hesitation as authentic by Professor Ritter of Berlin, and the able English geographer of Africa, Captain Macqueen.*

The following particulars were taken from Dr. Krapf's relation:—

---

greater reward" (p. 10). Dr. Beke says in the following page, "The word Doko is not to be regarded as the proper name of any particular people. In the Galla language it is an appellation, signifying ignorant, stupid; and it appears to be used in the same indefinite sense as our expression, savage." If the word Doko has a specific meaning, and is really the designation of a particular tribe, a quotation from a letter of M. d'Abbadie, given by Dr. Beke from the Athenæum of 8th March, 1845, p. 243, clearly demolishes Dilbo's account of the physical features of the Dokos: M. d'Abbadie says, "My Sidama interpreter was a Dokko, freed by his master's death. This man remained nearly two years with me, and was eighteen centimetres [seven English inches] shorter than myself. I have seen three other Dokko, all black like negroes, but with a fine facial angle like the Mozambique natives, and rather small—what we call *trapu* in France—but nothing like pygmies."—Ed.

* When Dr. Krapf was in S.E. Africa ($5\frac{1}{2}$° S. lat. near the coast) in 1848, he was informed by his native guide, that when he was in the Jagga country, he had seen little men from three and a half to four feet in height, who came down from a country to the N.W. selling iron in exchange for beads. This account seems to corroborate the story of the Dokos, so far as to the existence of a race of pygmies, without the strange circumstances told by Dilbo. The position of the country N.W. of the Jaggas would be that described as S.W. from Kaffa. Their Kaffir name, Wabilikimo, is merely descriptive, meaning "two cubits;" the Jaggas call them Koningo, a name almost identical with Koniunku, that of a large tribe of blacks (Kafirs) who had been brought down to the Cape in 1845, and were described by M. Arbousset, of the Paris Missionary Society, in the "Commermercial Advertiser," published at Cape Town. These came from a very remote part of the interior, some of them having been three or four months on their journey to Mozambique. But M. Arbousset makes no remark as to their stature, and he speaks in praise of their language as soft and musical, with pretty regular alternation of vowel and consonant. That there is a short race of men in the interior is not at all unlikely; physical causes have dwarfed several tribes of mankind in various regions; but when the men

"The countries on the west and south-west of Kaffa are, according to Dilbo, Dambaro, Bonga, Koolloo, Kootcha, Soosa, Toofte, and Doko; on the east and south-east are the plains Woratto, Walamo, and Talda.

"The country of Doko is a month's journey distant from Kaffa, and it seems that only those merchants who are dealers in slaves go farther than Kaffa. The most common route passes Kaffa in a south-westerly direction, leading to Dambaro, afterwards to Kootcha, Koolloo, and then passing the river Omo to Toofte, where they begin to hunt the slaves in Doko, of which chase I shall give a description as it has been stated to me, and the reader may use his own judgment respecting it.

"Dilbo begins with stating that the people of Doko, both men and women, are said to be not taller than boys nine or ten years old. They never exceed that height, even in the most advanced age. They go quite naked; their principal food is ants, snakes, mice, and other things which commonly are not used as food. They are said to be so skilful in finding out the ants and snakes, that Dilbo could not refrain from praising them greatly on that account. They are so fond of this food, that even when they have become acquainted with better aliment in Enarea and Kaffa, they are nevertheless frequently punished for following their inclination of digging in search of ants and snakes as soon as they are out of sight of their masters. The skins of snakes are worn by them about their necks, as ornaments. They also climb trees with great skill to fetch down the fruits, and in doing this they stretch their hands downwards and their legs upwards. They live in extensive forests of bamboo and

---

themselves come under the eye of an intelligent European observer, we usually find the marvellous portion of the account dwindle to very small dimensions.—ED.

Since writing the above note, the Editor is informed by Dr. Bleek that the Koniunku of M. Arbousset are the Niungwe, who reside in and about Tette, on the upper course of the Quilimane, or Zambezi River.

other woods, which are so thick that the slave-hunter finds it very difficult to follow them in these retreats. These hunters sometimes discover a great number of the Dokos sitting on the trees, and then use the artifice of showing them shining things, by which they are enticed to descend, when they are captured without difficulty. As soon as a Doko begins to cry he is killed, from the apprehension that this, as a sign of danger, will cause the others to take to their heels. Even the women climb on the trees, where in a few minutes a great number of them may be captured and sold into slavery.

"The Dokos live mixed together; men and women unite and separate as they please; and this Dilbo considers as the reason why that tribe has not been exterminated, though frequently a single slave-dealer returns home with a thousand of them reduced to slavery. The mother suckles the child only as long as she is unable to find ants and snakes for its food. She abandons it as soon as it can get its food by itself. No rank or order exists among the Dokos. Nobody orders, nobody obeys, nobody defends the country, nobody cares for the welfare of the nation. They make no attempt to secure themselves but by running away. They are as quick as monkeys; and they are very sensible of the misery prepared for them by the slave-hunters, who so frequently encircle their forests and drive them from thence into the open plains like beasts. When there pressed, they are often heard praying. They put their heads on the ground, and stretch their legs upwards, and cry, in a pitiful manner, 'Yer! yer!'* Thus they call on the Supreme Being, of whom they have some notion, and are said to exclaim, 'If you do exist, why do you suffer us to die, who do not ask for food or clothes, and who live on snakes, ants, and mice?' Dilbo stated that it was no rare thing to find five or six Dokos in such a position and state of mind. Sometimes these people

* The word Yéro, in the language of Kaffa, means God. See Dr. Beke's Vocabularies, in the Philological Society's Journal, vol. ii. 1845, p. 97.—ED.

quarrel among themselves, when they eat the fruit of the trees. Then the stronger one throws the weaker to the ground, and the latter is thus frequently killed in a miserable way.

"In their country it rains almost incessantly, at least from May to January; and even later the rain does not cease entirely. The climate is not cold, but very wet. The traveller, in going from Kaffa to Doko, must pass over a high country, and cross several rivers, which fall into the Gochob.

"The language of the Dokos is a kind of murmuring, which is understood by no one but themselves and their hunters. The Dokos evince much sense and skill in managing the affairs of their masters, to whom they are soon much attached, and they render themselves valuable to such a degree, that no native of Kaffa ever sells one of them to be sent out of the country, as Captain Clapperton says of the slaves of Nyffie. The very slaves of this people are in great request, and when once obtained are never again sold out of the country. The inhabitants of Enarea and Kaffa sell only those slaves which they have taken in their border-wars with the tribes living near them, but never a Doko. The Doko is also averse to being sold; he prefers death to separating from his master, to whom he has attached himself.

"The access to the country of Doko is very difficult, as the inhabitants of Dambaro, Koolloo, and Toofte, are enemies to the traders from Kaffa, though these tribes are dependent on Kaffa, and pay a tribute to its sovereigns: for these tribes are intent on preserving for themselves alone the exclusive privilege of hunting the Dokos, and of trading with the slaves thus obtained. Dilbo did not know whether the tribes residing south and west of the Dokos persecute this unhappy nation in the same cruel way.

"This is Dilbo's account of the Dokos, a nation of pygmies, who are found in so degraded a condition of human nature that it is difficult to give implicit credit to his account. The notion of a nation of pygmies in the interior of Africa is very ancient, as Herodotus speaks of them in II. 32."*

## CHAPTER XIV.

### OF THE RACE OF GALLA.

THE Galla are a race extensively spread in eastern intertropical Africa, who have become, during the last century, very formidable by their numbers, and threaten to overwhelm the Abyssinian empire. They are one of those races whose physical character—I mean the genuine and prevailing type of their bodily conformation—holds an intermediate place between the Arabian on one side, and the Negro on the other. We had very little correct information respecting the Galla till of late, and our knowledge of their history is still imperfect. I shall at present collect some leading facts from the best sources of information within my reach.

It appears, as observed on a former occasion,† that the Gallas are originally of the same family with two other widely-extended races of people in the countries to the southward of Abyssinia. Vocabularies of the languages of the Somali and of the Danákil give strong indication of affinity between these languages and that of the Galla. The Somali are, indeed, much changed by their conversion to Mohammedanism, and by commercial intercourse with

* Respecting the evident mixing up of monkeys and men in this account, the very valuable treatise of Mr. Ehrenberg may be read, which is printed in the "Memoirs of the Academy of Sciences" in Berlin (January 1833), under the title "Essay on the Cynocephalus of the Egyptians." K. Ritter.

† "Physical History of Mankind," vol. ii. p. 162 *et seq.*

other nations, to which their situation on the coast of the Indian Ocean, and in a region long the seat of traffic, has given rise, while the Gallas in the interior retain their primitive habits unchanged.*

The race of Galla are said to have among themselves a common national appellative, belonging to all the tribes: Oroma† is this name; that of Galla means, in their language, "invaders." Their national tradition imports that their patriarch Wolláboo came from beyond Bargámo, "the Great Water," and that his children were nine—Aroosi, Karaigo, Jillé, Abitchu, Ghelan, Wóberi, Metta, Gambitchu, and Betcho-Fugook. From these tribes descended the numerous clans which now form the principal population of inter-tropical Africa.‡

The Gallas in their native countries are an equestrian nomadic people, wealthy in flocks and herds, who roam over boundless pastures in the highlands of Africa. They are said also to cultivate fields of corn in rich valleys which intersect these mountains. They dwell in conically thatched cabins, like most of the inhabitants of African highlands. The women are celebrated for their beauty; they clothe in cotton garments, and smear themselves with oil; they braid their hair into an infinite number of minute tresses, which fall over their shoulders after the manner of the ancient Egyptians.

The present possessions of the Galla extend round Abyssinia to the west, east, and south-east, and enter between the snowy downs of Shoa and of Gondar. They border on the country of the Dankáli, upon Harrar, the Somáli, and the countries of Zendjero, Gurãgue, Caffa, and

---

* The Somali Grammar of Lieut. Rigby clearly proves, in spite of a good deal of glossarial difference, the connection of the Somali and Galla languages. See Lieut. Rigby's Somali Grammar and Vocabulary, in the ninth volume of the "Bombay Geographical Society's Transactions."—Ed.

† Orma signifies "Men" in the Galla language. See Krapf's Vocabulary.—Ed.

‡ Major Harris.

Narea. The extent of the Galla country is unknown. It seems to be mostly table-land, with a productive soil, and a mild and healthy climate. It is known to contain mountains which are inferior in elevation to the mountains of Shoa: they are inhabited by the Ittoo, the Alla, and other Galla tribes.

The Galla nation consists of numerous tribes. While in Shoa, M. Isenberg got the names of upwards of fifty of them, who are mostly near that country: there are others unknown.* All these tribes are independent of each other, but united by the same origin and the same language. Female government, according to ancient Ethiopian custom, exists among them: the tribe Moolofallada is ruled by a queen, named Tohāmĕ, who is a woman of warlike spirit. Among the eastern Galla tribes, a kind of patriarchal government exists.

M. Krapf says, that of forty Galla tribes known by name, the greater number are now tributary to the King of Shoa.† The account which he gives of the Gallas may be understood to apply to those tribes settled on the borders of Abyssinia, rather than to the great body of the nation farther southward. It differs, at least, from the statements of Major Harris. Krapf says that the Gallas have no priest, and that the sacrifices of cows or sheep, which they make on particular occasions to Wāk their divinity, are merely free-will offerings.

Some tribes have embraced Islàm; most of them still adhere to the ancient African paganism. "Their religion," says Isenberg, "resembles that of the Kafirs. They worship a Supreme Being, termed by them Wāk, whose priests, called Kalitshas, go about carrying a whip and bell with them, like the public fools, or Zekārotsh, in Tugray, and with the intestines of goats twisted round their necks,

* Their names are given by Isenberg: it would be useless to my readers were I to insert in these pages a long list of their barbarous appellations.

† Shoa is the most southerly of the states into which the Abyssinian empire has been divided: its inhabitants are Amharas.

making portentous gestures, and uttering unintelligible sounds. Like the Shamanists of the Siberians, and the consecrated orders of more illustrious nations, they are wizards, conjurers, soothsayers, augurs, haruspices, and physicians. Like the ancient Greeks, and Etruscans, and Romans, they divine by inspecting the entrails of goats. Occasionally, not regularly, the Galla pray to Wāk, and expect from him the accomplishment of their benedictions and anathemas. They have no distinct idea what Wāk is, but to his priests he reveals himself in dreams. Their oaths are characteristic; they sit down upon a pit covered with a hide, and imprecate upon themselves that, if they do not perform their vows, they may fall into such a pit. They have funeral ceremonies, and believe in a future state, which is one of moral retribution. Their occupations are agriculture, as well as pasturage: the art of forging metal is known to them, and much practised."

Though the Galla tribes are so much divided, they have a certain point of union among them. Gallas from all quarters perform pilgrimages to a certain tree called Wodanābè, situated on the banks of the Hawāsh, south of Shoa. This sacred tree is worshipped and addressed by prayers for riches, health, life, and every blessing. It is never approached by women.

According to Major Harris, the Gallas have two orders of priests: the Lúbahs, who perform sacrifices, and the Kalichas, or sorcerers. The Lúbah conducts the sacrifices of goats, offered twice in the year to the deities Ogli and Atéte; on which occasion, a number of goats having been slain, the Lúbah, wearing a tuft of long hair on the crown of his head, proceeds, with a bell in his hand, and his brows encircled with a fillet of copper, to divine from the fat, caul, and entrails, what success will befall the warriors of his tribe: after which the assembly, howling and screaming, surfeit themselves with raw meat and beer and tobacco-smoke, invoking Wāk, the supreme, to grant them long life, and Atéte, the goddess of fecundity, for her favours. The Kalicha, or wizard, ornamented with the putrefying entrails of a goat about his neck, and armed with a bell and copper whip, expels devils from the possessed, after paying adoration to a serpent. No Amhara will kill either a Lúbah or a Kalicha, from superstitious dread of his dying curse; and even the Christians of Shoa call upon the Galla sorceresses to clear their haunted houses from evil spirits, which is done by incantations and by the blood of ginger-coloured hens and red he-goats.*

---

* A very curious paper by Dr. Beke, printed in the first volume of the "Friend of the African," 1844, p. 90, is decisive as to the Gallas having received some knowledge of Christianity at some former epoch, and, in fact, of their adoption of some of its external features; though any real acquaintance with its tenets which they may have had has gradually faded away, and left nothing but a few names in a form which shows the source whence they were derived; all being slightly altered from the Ethiopic language, as clearly shown by Dr. Beke. The creator of all is Maremma, the Virgin Mary; her son, who is nearly upon an equality with her, and at whose command heaven and earth will pass away, is called Balawold, in Ethiopic the "Festival of the Son," adopted as a proper name. Sanbata and Kedami, meaning the Sabbath and the day preceding, are also great gods. Selassi, the Trinity, is another, but less potent. Maddin may be Medhani, the Saviour. Mikael and Gabriel are obvious enough, Gergis is St. George, and Telkamot the great Abyssinian Saint Tekla Haimanot. Abbogeramfas is a corruption of the Ethiopic, Abbo Gebra Menfas, to whom so many churches are dedicated in Abyssinia. Dabilos and Sietan, malignant beings, are plainly Diabolos and Satanas, two names considered as two distinct persons. The Gallas have some notion that Maremma gives food and good things to good men after death, and that the wicked are burned in fire. They honour each of their great deities monthly, by fasting or feasting; Kedami and Sanbata are celebrated every week: they keep the festivals of the Assumption and Nativity every year, and hold a fast of twelve days preceding the former. Although the Gallas have neither priests nor churches, nor forms of prayer, each man making his petition to any god in any way and at any time as best suits himself, there can be little doubt that their religion, such as it is, and however deplorably fallen away from a model which was already in a lamentable state of degradation, retains sufficient traces of that model at least to point out its original source. The declension is perhaps what might have been expected among an illiterate and nomade race of men, whose teachers were not much more elevated than

*Physical Characters of the Gallas.*—Few of the travellers in Abyssinia have thought it worth while to give us any account of the physical character of the Gallas. Bruce merely says of them that they are of a brown complexion, with long black hair: he adds, that some, who live in the valleys of the low country, are perfectly black. Isenberg says that the men are not more handsome than the Abyssins, but that the women are prized chiefly for their light or fair complexion. Mr. Salt, in his last Travels, gave the portrait of an Edjow Galla, of which the following sketch is a copy.

FIG. 72.

Edjow Galla.

themselves, and who had no outward forms to aid in maintaining the inward spirit, no separate class to learn and to communicate the doctrines, nor any rules to enforce or persuade the observance, of the religion professed. It has thus become little better, either in theory or practice, than the fetichism and sorcery of the Negroes and Kafirs.—ED.

There are two or three portraits of Gallas in Lord Valentia's Travels which display much better the prevalent type of their countenance: they agree, at least, more closely with the description given of the people by Dr. Rüppell and by M. d'Abbadie. The portait of a Galla boy is from a drawing furnished by the latter traveller.

Dr. Rüppell has given a brief but characteristic description of their physical character, which is common, as he says, to more than one nation in the east of Africa, particularly to the Galla race and to the Shoho, or Hazorta, who have been already mentioned. He says, "Their countenance is rounder than that of other Abyssinian nations; their noses are straight, but short, and divided from their foreheads by a sinking-in (*eine Vertiefung*); their lips are rather thick, but yet not like those of Negroes; their hair is thick, strongly frizzled, and almost woolly (*beinahe wolliges*); their eyes are small, deeply set, but very lively; their persons are rather large and bulky."

It seems from this writer, that the Galla are one of those almost woolly-haired races, with round faces, obtuse and thick features, and thick lips, and dark complexions, who, in Eastern Africa, fill up the transition from the Syro-Arabian type to that of the Western Negro.

The Galla in the north may bear a comparison with the Kafir nations of South Africa, as to the extent of country which they have reduced under their sway, or have laid waste by their predatory incursions; and both may be compared, in this respect, to the Turks or Mongolians in Northern Asia. The Kafirs and the Gallas probably divide between them nearly all the vast extent of the Central African plateau, the former occupying the southern, and the latter the northern region. Both appear to be races of tall stature, and possessed of great physical energy. They differ otherwise in physical characters. The Kafirs, though they recede far from the genuine African or Negro type in the shape of their features and in the complexion of at least some tribes, yet belong to the class of woolly-

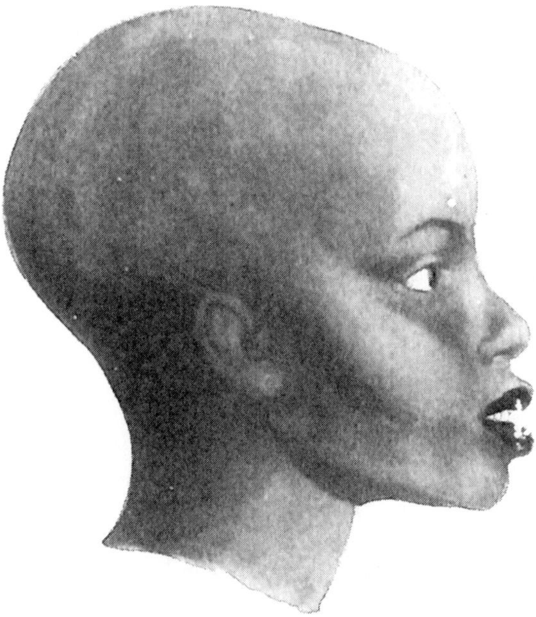

haired nations. The Galla, on the other hand, setting out (if we may use the expression) from the Asiatic or Syro-Arabian type, make but an approximation towards the characteristics of the African.

The Danákil call themselves Affar;* the name of Danákil is given them by the Arabs. Adaiel, a name very celebrated in these parts of Africa, also belongs to the race of Danákil. It was originally Ad Alli, one of the tribes of the Danákil race, who form the population of Tajura. Aussa is peopled by another Danákil tribe, called Mudaitu.

The race of Danákil long ago adopted Islàm, and their kingdom of Adel or Adaiel embraced the whole Mohammedan population of the eastern part of Africa.† Besides the great empire of Adel, we are informed that there is a particular province to which that name more strictly belongs, probably the original seat of the Ad Alli tribe. The kingdom of Adel was nearly destroyed by the invasion of Amda Sion, emperor of Abyssinia, between 1312 and 1342; and after that time Aden or Harrar became the seat of the chief power of the Moslims, and the capital of the kingdom of Adel. The Harrar people, however, would appear to be a different race. We are informed that their language is allied to the Amharic. They are probably a branch of the Amharic race who embraced Islàm and founded a kingdom ever in hostility with the Abyssinian monarchy.‡ But the history of Adel

* Ophir, Afri. The Danákil are also termed Ghibertis, a name frequently occurring in the works of writers on Abyssinia. Ghiberti means "strong in the faith."

† Macqueen, 20.

‡ Harrar is now a great commercial town, though of not so much importance as when it was the centre of the Adaiely monarchy. The country is surrounded by Galla tribes. The Harrari are rigid Moslims. Though their language bears an affinity, generally recognised, with the Amharic, they are said to write in the Arabian character. This is, indeed, not unlikely, since the Mohammedans of Shoa write the Amharic language in that character. See "Report on the Geographical Position of Harrar, and

is connected, in later times, with the fate of Abyssinia. From the time of Amda Sion to the fifteenth century, Adel was subject to the Negus; but in the year 1528, Mohammed Gragne,* king of Adel, who reigned in Harrar, conquered Shoa, and laid waste the whole country of the Amhara, and so weakened the power of the Abyssinian monarchs, that they were unable to resist the incursions of the ferocious Pagan Gallas, which took place soon afterwards.

Through the small kingdom of Bali, to the eastward of Efat, the Gallas first rushed into Abyssinia, in 1559. They conquered all the southern parts of the empire, and would probably have entirely destroyed it, had they not weakened their power, at length, by wars among themselves, and given an opportunity to the Abyssinian chieftains to rally, and gradually reduce them under their power.

The Gallas (though, as we have said, their language indicates affinity to the dialects of the Adaiel and the Somáli) appear in history as a distinct and entirely new people in the age when they suddenly rendered themselves formidable.

It would be a very important result in the ethnology of Africa, and in that of the Syro-Arabian family of nations, if it should be established that the language of the Galla, originally cognate as it seems to be with the dialects of so many widely-spread Ethiopian races, is one of those Hebræo-African languages of which the existence has been found among the dispersed nations between the Red Sea and the Atlantic Ocean. The specimens hitherto published of the Galla language are, perhaps, too few to authorise a very positive conclusion; but they afford most striking examples of analogy, and I think there is little

on the Tribes in its Vicinity," by Lieut.-Col. A. Barker, attached to the mission to Shoa; "Geographical Journal," vol. ii. p. 238. Major Harris gives the same statement as to the affinity of the language of the Harrari with the Amharic.

* Gragne means left-handed.

room for doubt, that the Galla speech will turn out to be a kindred dialect, or, perhaps, rather a sister-language of the Ghíz, the Amharic, and the Berber idioms.*

## CHAPTER XV.

### OF THE SOUDANIAN RACES, OR BLACK NATIONS INHABITING THE INTERIOR OF AFRICA.

NOTHING has tended more to spread vague and false notions in subjects connected with ethnology than the improper use of general names. It is often made a question, what races among the inhabitants of Africa are Negroes, the meaning of the term not being defined. It ought to be remembered that the word Negro is not a national appellation, but denotes the ideal type constituted by the assemblage of certain physical characters, which is exemplified in the natives of Guinea in Western Africa, and in their descendants in America and the West Indies. When these characteristics are not all found, it has often been said that African nations, though black, or nearly black, and woolly-haired, are not Negroes. Thus the Kafirs and Hottentots are said not to be Negroes. On the same principle, we ought to except the nations of the interior of Africa or of Soudan, in some of whom we could scarcely recognise any considerable resemblance in features to the Negroes of Guinea.

Central Africa is supposed to be divided by a vast transverse range of mountains, which runs across the whole Continent, about ten degrees to the northward of

* In M. Krapf's short Grammar of the Galla, Mr. Newman, whose analysis of the Berber grammar is well known, has pointed out numerous and extensive analogies between these two languages. The Tuaryk probably border on the Galla towards the east, and there is great probability in the conjecture that both originally belonged to one family.

the equator, from Cape Guardafui, on the eastern, to Cape Roxo, on the western side. A part of this chain, towards the east, was termed by the ancients the Mountains of the Moon, where the Nile was supposed to take its rise. The western part, above Mandara, as we are assured by Denham and Clapperton, is now called by the Mohammedans Jebel-Kumra, which has the same meaning; and the whole chain, the continuity of which is, however, rather probable than fully proved, receives from modern geographers the like appellation. This chain separates the comparatively civilised region containing the Mohammedan states, or empires of Africa, from the vast and unknown wilderness to the southward, from which camels and caravans—the ships and fleets of the desert—are excluded. The chain of Kong, traversing in a similar direction the great western projecting part of Africa, is supposed to be a prolongation of the same system of mountains. It is immediately to the southward of this chain that those African races exist whose aspect displays the characteristics of the Negroes fully developed, and in the highest degree.

The mountains of Mandara, according to Denham, are not of great elevation, but they are only the outskirts of a vast Alpine chain. They were asserted to extend southward a journey of two months, and in some places to be ten times as high as those which rise above the plains of Mandara. The only communication with the region lying farther towards the south is by means of a few adventurous freed slaves, who penetrate into the interior of the mountainous tracts with beads and other articles of traffic from Soudan, slaves and skins being given in exchange. The nations who inhabit this wilderness are very numerous. They generally paint and stain their bodies of different colours, and live in common without any regard to relationship. Large lakes are frequently met with in this country plentifully supplied with fish. Mangoes, wild figs, and ground-nuts, are found in the valleys. The people of these mountains, seen by Denham, are described as having their heads covered with long woolly, or rather bristly hair, coming down over their eyes; round their arms, and in their ears, were rings of what appeared to be bone, and around the necks of each were from one to six strings of the teeth of the enemies they had slain in battle; teeth and pieces of bone were also pendant from the clotted locks of their hair; their bodies were marked in different places with red patches, and their teeth were stained of the same colour. Their whole appearance is said to have been strikingly wild and truly savage. Endeavours to set on foot intercourse with them were in vain,—they would hold no communication; but, having obtained leave, carried off the carcass of a horse to the mountains, where the fires that blazed during the night, and the savage yells that reached the valley, proved that they were celebrating their brutal feast.*

To the northward of the line above marked out, the nations of Africa are comparatively civilised. They practise agriculture, and have learned the necessary, and even

* This account is given by Major Denham of the Musgow people; but Dr. Prichard has omitted the singular close of the paragraph from which the above account is taken, probably on account of its apparent absurdity; yet the notice of the religion of the Gallas, in p. 313, *ante*, shows that it might not have been wholly groundless. Denham's words are: "What very much increased the interest I felt in gazing upon these beings, who, to appearance, were the most savage of their race, was the positive assertion of Boo-khaloom that they were Christians. I had certainly no other argument at the moment to use, in refutation of his position, but their most unchristianlike appearance and deportment; in this he agreed, but added, 'Wolla Insara,' they are Christians!"

The travellers Barth and Overweg, at the beginning of 1852, accompanied the Bornu army on a slaving-expedition, in the same way as Denham had done twenty years previously. They found a dense and industrious agricultural population, numerous groups of habitations, and fields well and carefully cultivated. They write that the Musgow people are well-made, though not handsome; and that the women's faces are horribly disfigured by an ornament of ivory worn in the upper lip. They saw some hills of moderate elevation, but no mountains. (Petermann's Account of the Progress of the Expedition to Central Africa, 1854, p. 9.)—Ed.

some of the ornamental arts of life, and dwell in towns of considerable extent, many of which are said to contain 10,000, and even 30,000 inhabitants,—a circumstance which implies a considerable advancement in industry and the resources of subsistence. All these improvements were introduced into the interior of Africa three or four centuries ago; and we have historical testimony that, in the region where trade and agriculture now prevail, the population consisted, previous to the introduction of Islàm, of savages as wild and fierce as the natives farther toward the south, whither the missionaries of that religion have never penetrated. It hence appears that human society has not been, in all parts of Africa, stationary and unprogressive from age to age.

In Mr. Park's account of Sego, the capital of Bambarra, which contains about 30,000 inhabitants, the houses have two stories, and flat roofs: mosques are seen in every quarter, and ferries conveying men and horses over the Niger. "The view of this extensive city," says Mr. Park, "the numerous canoes upon the river, the crowded population, and the cultivated state of the surrounding country, formed altogether a prospect of civilisation and magnificence which I little expected to find in the bosom of Africa." To the eastward, he passed a large town called Kabba,—situated, as he says, in the middle of a beautiful and highly cultivated country, bearing a greater resemblance to the centre of England than to what he should have supposed to exist in the centre of Africa.

The earliest accounts of Soudan are to be found in the works of Arabian geographers and travellers, of whom Edrisi, Ibn Batuta, and Leo Africanus, are the most celebrated. In the works of the latter we find described nearly all the countries known in the interior of Africa to the northward of the great chain of mountains. It seems that Mohammedan states had been recently founded in this region. Leo, who appears to have had a better notion of what was required for ethnology than any of the Greeks or Romans, has informed us what races and languages extended through each great district. He divides the whole interior of Africa known to him into fifteen states, which in his time were subject to four kings; and he intimates that through each of these kingdoms a particular race and a distinct language was extended. The princes of these states were Mohammedans. Omar and Abraham, or, perhaps, Ibrahim, were among their names. The eastern kingdom was Gaoga, comprehending countries to the eastward of Bornu. Bornu was the second; Guber, now part of Haussa, the third; and Tombutum, the western state.

We have some short specimens of the idioms prevalent among these nations of interior Soudan, which, though not sufficient for the foundation of any positive conclusion, yet give reason to suspect they are all referable to one great family of languages.* The physical characters

* The vocabularies of nearly 200 languages have very recently been brought from Africa by the Rev. Dr. Koelle, and are now being printed. They are all collected on a uniform plan, and they contain sentences, as well as distinct words, by which they are made very much more valuable for philological comparison, enabling the investigator to collate something of structure as well as glossary. A slight examination of these vocabularies seems to show that there are among the Negro idioms a dozen or more classes of languages, differing from each other at least as much as the more remote Indo-germanic languages do. We are unable to say more at present, and the most careful examination of all we have might be insufficient to enable us to pronounce with safety: limited vocabularies, although aided by such a knowledge of structure as could be obtained from short sentences, would hardly enable us to pronounce upon the alliance of Sanskrit, Slavonic, Latin, and German, which we know to be connected. It is still more difficult to decide as to radical difference: our materials are often sufficient to show a connection, but are rarely extensive enough to prove the negative proposition. The opinion of Dr. Bleek, that most if not all the West African idioms are allied to the great South African or Kaffir language, is undeniably correct in the main; the Editor can speak positively to this for several of those languages which show a like disintegration of Kaffir as is seen in French and English compared with Sanscrit. This, of course, does not involve the superior antiquity of the Kaffirs, rather, perhaps, the contrary; the emigrating populations who passed the Equator would be less exposed to influences producing change than their relations left behind them. The same phenomenon is exhibited in Iceland, where a race of Scandinavian colonists has preserved the ancient language which the older communities of Sweden, Norway, and Denmark, have lost.—ED.

of the people differ considerably. The Bornawi are reported by all travellers to be more like the ideal Negro than the natives of Haussa. The latter are described as handsome people, with woolly hair, and a jet black complexion. The specimen of the race given in Plate xiii. was taken from a native in London. The features are remarkably good, and appear to have little resemblance to those of the natives of Guinea. Mr. Jackson assures us that they are acute, intelligent, and industrious. "They possess a peculiarly open and noble countenance, having prominent noses and expressive black eyes."

That such a people should betray *any* relationship to the Shemite nations is a fact so contrary to prejudice and prevailing opinion, that the assertion will not be believed without proofs, and these could not well be displayed within our limited space. For the sake of those who feel curious upon the subject, I must beg to refer to an appendix to the fourth volume of my "Physical History of Mankind," written by Mr. F. W. Newman, from materials furnished by M. Schön.*

## SENEGAMBIAN NATIONS.

By the name of Senegambia is designated that part of Africa which contains the rivers Gambia and Senegal. It lies to the northward of the Kong, which, as I have observed, is a western prolongation of the great chain traversing the African continent from east to west. The whole of this high region is rather a great plateau, or table-land, than a chain of hills; it presents three lofty

* An essay on Bornu Grammar, written by the Editor from some materials sent to England by the late Mr. Richardson, and still more the large Bornu grammar since published by Dr. Koelle, show a grammatical structure quite different from anything Semitic, and equally so from any other Negro tongue known. The vocabulary contains many Arabic words, mostly borrowed, but some which can hardly be so; it is generally different from that of other Negro tongues, though we see several Tibbu words, and here and there a Haussa or Mandara word may be detected.—ED.

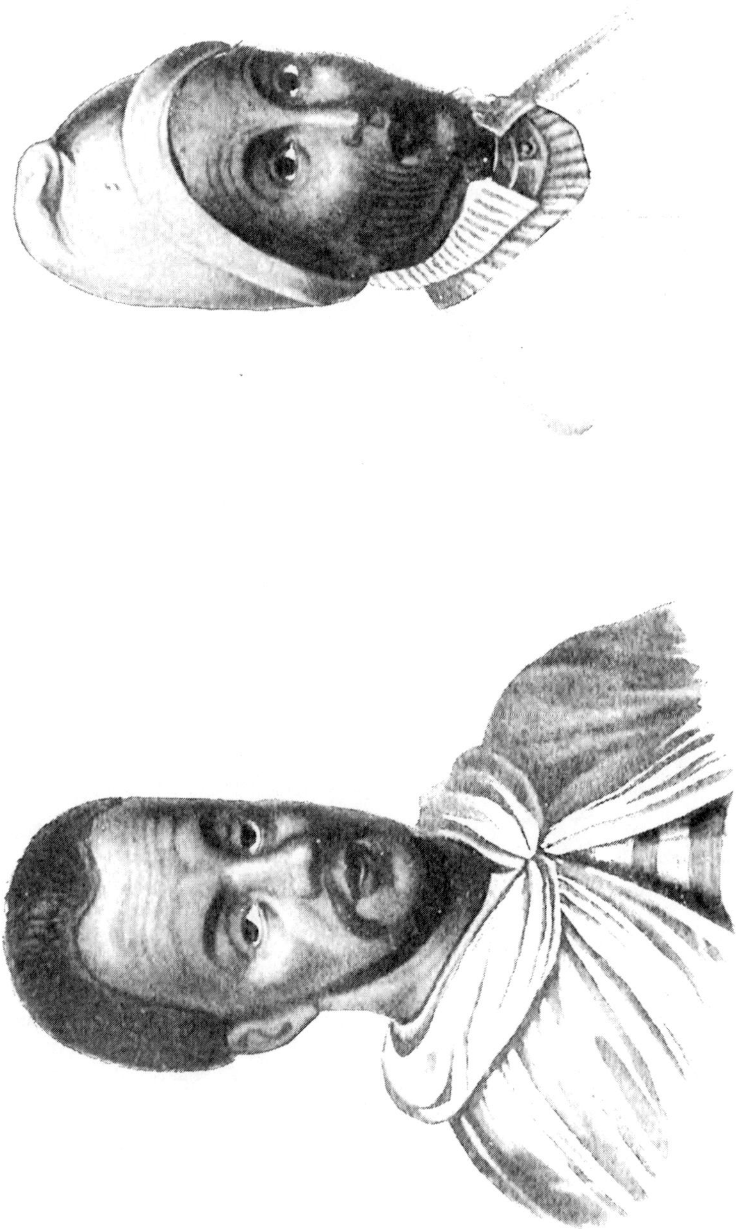

fronts towards the sea and the surrounding low countries, consisting of high terrasses, or mountainous uplands. The northern declivity contains long tracts of fertile country, which, with the valleys belonging to it, and receiving and conveying its great rivers towards the ocean, forms the native land of one of the most powerful, and numerous, and intelligent of the African races, namely the Mandingos. The western front, fanned by the breezes which blow from the Atlantic Ocean, and in some parts cold from the great elevation of the surface, is, in the opinion of the most learned of modern geographers,* the father-land or primitive dwelling of the race of Fulahs. The southern front of the mountains of Kong, with the lowlands underlying it, and the border of sea-coast, is Guinea, the region of genuine Negroes, where the peculiar features and physical and moral qualities of the Negro races are developed in the highest degree.

I shall describe in this section the Senegambian races, namely the Mandingos, the Iolofs, the Fulahs, and some other tribes inhabiting the neighbouring countries: in the following section I shall advert to the Negroes of Guinea.

1. *The Mandingos.*†—The Mandingos are remarkable among the nations of Africa for their industry; and of all the intertropical races of that continent, they have evinced the greatest energy of character. They are the most zealous disciples of Islam, and abstain from all intoxicating liquors. "The Mandingo merchants," says Golberry, an intelligent French traveller, "among whom are many marabouts, or priests, are men of enterprise and intelligence." They possess great influence in Northern Africa, and carry on the principal traffic in that country. The Mandingos are active and shrewd merchants, laborious and industrious agriculturists; they keep their ground well

* Professor Karl Ritter.

† In my "Researches into the Physical History of Mankind," vol. ii. on African ethnography, I have entered fully into the history of the Mandingo race.

cultivated, and breed a good stock of cattle, oxen, sheep, and goats. They are a kind and hospitable people.

The colour of the Mandingos is black, with a mixture of yellow: in this they contrast themselves with the Fulah, whose colour is red mixed with yellow. Golberry declares that, in their features, they resemble the Blacks of India more than those of Africa. "Their features are regular, their character generous and open, and their manners gentle. Their hair is of the kind termed completely woolly." Park says they are not so handsome as the Iolofs, who are the most beautiful, and at the same time the blackest people in Africa. "The women have the management of domestic affairs, are very cheerful and frank in their behaviour, and instances of conjugal infidelity are rare."

It appears probable, from reasons which I have elsewhere stated, that the people of Iallonkadou, above the Mandingo country, are a branch of the same race, as well as the other nations of the highlands behind Cape Verd and Sierra Leone.* Among them are the Sulimanians, a warlike people, who, according to Major Laing, resemble in many of their customs the ancient Romans.

In the low countries toward Cape Verd is the territory of the Bourb' Iolof, or Iolof emperor. The Iolofs have been known since the fifteenth century: they are a people of mild and social character, and are described as remarkably beautiful. Their complexion is a "fine, transparent, deep black; their features are like those of Europeans, except that their lips are rather thick."†

---

* A manuscript vocabulary of the Jallonkadu idiom recently seen by the Editor is almost wholly Susú.—Ed.

† Sixteen tribes or nations supposed to be connected in language with the Mandingoes are provisionally given by Dr. R. G. Latham at p. 473 of his "Varieties of Man," and more will be found to result from a collation of the Rev. Dr. Koelle's vocabularies. One of these, the Vais or Veis, a people whose habitat reaches the coast at the Gallinas, is peculiarly interesting as possessing a native alphabet or syllabarium of recent invention, with which they have written several books: three of them have been brought to

2. *The Fulahs.*—The Fulahs are one of the most remarkable nations in Africa, and their origin is a subject of great interest. They have long been known to traders in Western Africa, and by all the old writers have been included among the Negro nations. By De Barros, the mountainous country near the source of the Rio Grande was pointed out as the kingdom of Temala, sovereign of the Fouli, who reigned there in 1534, and carried on war with Mandi Mansa, *i.e.* Mandingo King. On the border of Senegambia, about the sources of the Rio Grande, and on the slope, or terrass, which looks towards the setting sun, and is cooled by the higher currents of air flowing from the Atlantic, are the elevated plains inhabited by the Fulahs. Timbú, their capital,—like ancient Rome, a military station, or centre of conquests—contains 9000 inhabitants: it is surrounded in part by dry and rocky deserts, and partly by mountain pastures, which feed numerous flocks of sheep and goats, and herds of oxen and horses, unknown in the lower regions. The inhabitants of this Alpine country, who differ physically from the natives of the lower region, cultivate their soil with industry; but such has been their seclusion from the rest of mankind, that the use of the plough is still to them unknown. They forge iron and silver, work skilfully with leather and wood, and fabricate cloth. They have clean and commodious dwellings, and have had mosques and schools in their towns since Islàm was introduced among them by marabouts from the Mandingos. Their armies are victorious over the neighbouring nations, and are said to have extended the dominion of Timbú over forty geographical miles from south to north, and seventy-eight from east to west. The sovereign,

---

Europe, and one has been published in London by lithography, and privately distributed. See the account of the discovery of this curious fact by Capt. Forbes, and of the grammatical notices of the language by the Editor of this work in the 20th volume of the "Journal of the Royal Geographical Society." It will be seen that the Timmani and Bullom must be withdrawn from the category of Mandingo languages.—Ed.

or the Almamy, of the Fulahs reigns at Timbú. His country, Fouta-diallo, contains other considerable towns, Temby and Laby, the capital of Cacoundy, a district well cultivated, and producing abundantly rice, oranges, and maize.

Fouta-diallo, or Fouta-jallo, is, however, but a part of the territory now occupied by the Fulahs in Africa. They are spread in various tribes over the countries between the Senegal and Gambia rivers, and in the region farther towards the south. According to M. Golberry, they constitute the most numerous part of the population from the fourth degree of northern latitude to the Senegal. One of the principal Fulah states, and that in which they became known from the earliest period to Europeans, is the kingdom of the Siratik, or Fulah sultan, on the Senegal, which includes an extensive territory on that river, reaching from the borders of Galam to Fort Podhor and the Lake of Cayor. In this country, the Fuli, or Pholeys, were visited by Jobson, Le Maire, and the Sieur de Brüe, in the seventeenth century, when the court of the Siratik is said to have displayed much barbaric magnificence. The fertile country of Bondu, near the sources of the Nerico, though subject to the conquering Mandingos, is likewise chiefly inhabited by Fulahs. The same people occupy a great part of Brouka, to the eastward of Bambouk, as well as Wasselah, on the higher course of the Niger. In the high countries on the eastern part of Senegambia, there is a mountainous tract near the source of the Senegal which bears the name of Fouladou, or Wilderness of the Fulahs. The inhabitants of that country are a wild and savage people. The name which their territory bears would seem to imply that it is looked upon as the original or proper habitation of the Fulah race.

The identity of race which connects the Fulah of Senegambia with the Felatahs of Central Africa was a discovery of Professor Vater. It is no longer a matter of doubt that these nations belong to the same stock: they have similar physical characters, and speak one language. The history of the Felatahs is very remarkable: their rise in Africa may almost be compared with that of the Arabs in the time of Mohammed. According to information obtained by the enterprising and much lamented Clapperton, the Felatahs wandered out originally from the country of Melli, under which term they include the Fulah states in Senegambia, Foota-Torro, Foota-Bondu, or Bondan, and Foota-Diallo. The wandering Felatahs, like the Fulah hordes in the borders of the Iolofs, lived, as we have observed, in forests, and fed cattle. They dispersed themselves over the greater part of Soudan, and, being everywhere disregarded and despised, their numbers were unknown. Many hordes still continued to be Pagans, but those who had embraced Islam became devotees and zealots for their religion: they performed the pilgrimage to Mecca; many also visited the cities in Barbary. They increased in intelligence, but never formed themselves into a nation, until a revolution took place in their habits and character, parallel in many respects to the change induced among the Arabs at the first outbreaking of the Mohammedan enthusiasm. The author of this revolution was a Felatah sheikh, named Othman, commonly termed Danfodio, who acquired all the learning of the Arabs in Africa, and succeeded in persuading his countrymen that he was a prophet. Having laid this foundation of his power, he came out of the woods of Ader, or Tadela, and built a town in the province of Guber, where the Felatahs gathered round him. Being expelled by the people of Guber, Danfodio, with his Felatah followers, returned to Ader, and built a town which they called Saccatú. To the people of his race, who flocked to him from different countries, he gave different chiefs, telling them to go and conquer in the name of God and the Prophet, who had given the Felatahs the lands and all the riches of the Kafirs. Each chief bore a white flag; the Felatahs were to wear white robes, emblems of their purity; and their war-cry was to be Allah Akbar. Their confidence in the supernatural power of their chief inspired them with valour. They conquered Kano without a blow, overran Guber, and killed the sultan: they subdued afterwards the whole of Haussa, with Cubbe, Youri, and a part of Nyffé; they attacked Bornu on the east, and Yarriba on the west, of which they conquered a part; and once entered the capital city, Eyeo or Katunga. Danfodio was an object of terror among all the Negro nations in the interior. Some years before his death, Danfodio became religiously mad; but until that time his government was well regulated. At his death, in the year of the Hegira 1232 (1816-7), Guber, Zamfra, a part of Kashna, and Zegzeg, threw off the yoke of the Felatahs; but the chieftain of Saccatú, Mohammed Bello, succeeded in reducing a great part of the country under his dominion.

Similar accounts of the progress of the Felatahs were given to Mr. Lander, who, in his passage through different Negro states, collected many additional particulars relative to the conquests and dispersion of that people. He says that the Felatahs in former times never resided in towns, but wandered with their flocks and herds in small companies. They stole into Haussa imperceptibly, and were at length so numerous in that country as to be enabled to form a powerful combination for its conquest, and the establishment of their own empire of Saccatú. Most of the Felatahs are Moslemin, but many hordes are still Pagans: both Clapperton and Lander declare that these are precisely the same people in other respects, that they have exactly the same language, and the same features and complexion. Lander says that they have been dispersed over the Borgú territory from time immemorial. The Felatahs, in Borgú, maintain no intercourse with people of their own kindred in Haussa, where they are the dominant race, nor have they the slightest idea or tradition of their origin. They are generally termed Foulanie, and speak, as Lander says, the same language, and follow the same pursuits, as the Fulahs near Sierra Leone.

The Fulahs are described by M. Golberry as "fine men, robust and courageous. They have a strong mind, and are mysterious and prudent. They understand commerce, and travel in the capacity of merchants, even to the extent of the Gulf of Guinea. They are formidable to their neighbours. Their women are handsome and sprightly. The colour of their skin is a kind of reddish black; their countenances are regular, and their hair is longer and not so woolly as that of the common Negroes; their language is altogether different from that of the nations by whom they are surrounded—it is more elegant and sonorous."

The tribes of Fulahs which, under the name of Poules, or Peuls, have peopled the borders of the Senegal between Podhor and Galam, are black, with a tinge of red or copper colour: they are, in general, handsome and well made. The women are handsome, but proud and indolent.

The enterprising traveller, Richard Lander, who had been among the Kafirs near Graham's Town in South Africa, before he visited the country of the Felatahs, was struck by the resemblance which these tribes of people bear to each other; and he confidently expressed an opinion that they are of the same race. He describes the Felatahs near Borgú as differing little in feature or in colour from the Negroes, but as having much longer hair, which they weave on both sides of the head into queues, and tie under the chin. This do many of the long woolly-haired nations of Western Africa. The late Captain Allen confirmed this account; he has assured me that the Fulahs, whom he had been accustomed to see near the Quorra, were not of much lighter complexion than the Negroes, and that he considered the difference between these races to have been represented as greater than it is. Nor was the assimilation, in his opinion, attributable to mixture, their connexions being chiefly in their own tribe. There must be a great difference between the different nations of this race; but the real nature and cause of this diversity remains to be discovered.

The able author of an elaborate memoir on the history of the Fulah race maintains the singular opinion that the Fulahs, though long known as an African nation, is an offset of the Polynesian race. The evidence brought forward in proof of this opinion consists of some words of analogous sound in the Fulah and Polynesian languages. The subject well deserves a careful consideration; and the supposition of M. d'Eichthal, though at first sight it appears improbable, especially if we take into account the distant period from which the Fulahs are known in Africa, and the difference of physical characters and manners, yet ought not to be dismissed without careful investigation. If sufficient means were accessible for acquiring a complete knowledge of the Fulah speech, this question could be speedily elucidated. The instances of resemblance in these languages as yet discovered by the ingenious author of the essay are so few and so remote, that it appears to me very doubtful whether any conclusion whatever can be founded upon them. A greater number of words might be discovered common to languages which confessedly have no relations. The number of similar words in the Fulah and the Polynesian dialects is extremely small; and they are, moreover, gleaned from a vast number of languages, which, though of kindred origin, have great diversity in their vocabularies. If the writer had taken the whole aggregate of the European languages for the subject of comparison with any one idiom either of Africa or America, he might have discovered more numerous and more striking coincidences; and yet such a method of proceeding is not very different from that which M. d'Eichthal has adopted.* With all the deference that is due to so able and ingenious a writer as M. d'Eichthal is well known to be, I am still

* The following instances, not of affinity, but of coincidence in particular words, were pointed out by Professor Vater, between the Celtic language of Ireland and the idiom of the Algonquins in North America:—

of opinion that the Fulahs are a genuine African race, that their language has a form of words and of euphony which places it in near relation to the dialects of other Senegambian races, and that there will not be found to exist so decided a line of demarcation between them and the black Sudanian nations as it has been supposed.*

### NEGRO NATIONS OF GUINEA.

The true Negro characters are chiefly found, as I have observed, in the nations who live to the southward of the great chain of mountains which has its western termination at the Sierra, and in that region they are most strongly displayed on the sea-coast which encircles the projecting region of western Africa to the inmost angle of the Bight of Benin. Within this vast extent of maritime country there are some nations considerably advanced in civilisation, and these have rather the Soudanian than the true

---

| | Irish. | Algonquin. |
|---|---|---|
| Island | Inis | Inis |
| Lee | Gai | Ga |
| Water | Uisge | Isca |
| Soft | Boy | Boye |
| All | Cac'uile | Kak eli |
| Everything | Cac'cine | Kak ina |

These instances of resemblance are more striking than any discovered between the Fulah and the Polynesian languages.

In the third volume of the "Mithridates," we find a rather long list of words scarcely less strikingly similar in the idiom of the Araucans and the Greek and Latin. Something more than the evidence of a few particular words is called for before we can admit the supposition of a common origin between separated nations.

* The opinion of the author is shown to be correct by the short Fulah Grammar deposited in the British Museum by the Rev. R. M. Macbrair. A very limited number of copies of this Grammar, with some additions and a slightly changed arrangement, was printed a few months ago (June 1854) by the Editor of this work, for the use of the expedition preparing to explore Adamaua, and it is clear that the language has analogies with Yoruba, Accra, Ashanti, Timmani, and some other Negro tongues; but the glossarial affinities are very limited. Dr. Bleek has found a good many Fulah words in the Wolof language.—ED.

Negro characters; but the greater number are savages of the rudest class, and these display most strongly the prognathous form of head and countenance.

The Feloops, near the Casamanca, are naked savages, their colour is a deep black, and their skin rough; but they are said to have handsome features. The Papels, the Bisagos, the Balantes, the Biafares, or Iolas, on the coast, are savages of the ugliest description. The Susús and Timmaní are more civilised; their features are more like those which prevail among the nations of interior Soudan.*

From the river Assiní, or from Cape Tres Puntas, to the Rio Volta, is the extent of the Gold Coast. The Amina race and language prevail through this region, and reach to an unknown distance in the interior; the Fantí, the Ashantí, the Akwapim, and the Inta, being of the same stock. The following curious account is given of these people by Barbot:—"The blacks in this part of Guinea are generally well-limbed and proportioned, being neither of the biggest nor of the lowest size and stature; they have good oval faces, sparkling eyes, small ears, and their eye-

* The Susús are usually classed among the Mandingoes, and their language justifies this classification; but a short examination of the language of the Timmanís, who are also usually classed among the Mandingoes, and who occupy the country rather to the north of Sierra Leone, shows most decidedly that this idiom is akin to the great Kafir family of languages. The Editor some time since received portions of two chapters from the Gospels of Luke and John, printed in 1847, from which he learned this curious fact; but it has been pointed out to him that the connection is mentioned by the learned Bishop of Sierra Leone in his interesting and valuable introduction to Crowther's Yoruba Grammar, printed in 1852. The bishop says (page 7), "I have recently discovered the Temneh (with its two cognates, the Sherbro and the Bullom) to be a branch of the great South African language." Dr. Bleek also noticed this connection at page 32 of his Treatise on the Languages of Southern Africa, printed at Bonn in 1851. He has recently shown the Editor that the languages of Western Africa generally have traces of that marked structure which has been usually considered peculiarly Kaffir. In several of them this is obvious; in others not so certain: in the Bornu language the Editor can see no trace of affinity to Kaffir. See the note in p. 323.—ED.

brows lofty and thick; their mouths not too large; curious, clean, white, and well-arranged teeth; fresh red lips, not so thick and hanging down as those of Angola, nor their noses so broad. For the most part they have long curled hair, sometimes reaching down to their shoulders, and not so very coarse as theirs at Angola, and very little beards before they are thirty years of age. The elderly men wear their beards pretty long.

"Their skin, though but an indifferent black, is always sleek and smooth. Their stomach is naturally hot, capable of digesting the hardest meat, and even raw entrails of fowls, which many of them will eat very greedily. They take particular care to wash their whole bodies morning and evening, and anoint them all over with palm-oil, which they reckon wholesome, and that it preserves them from vermin, which they are naturally apt to breed. In short, they are for the most part well-set, handsome men in their outward appearance, but inwardly very vicious.

"The black women are straight, of moderate stature, and pretty plump, having small round heads, sparkling eyes, for the most part *high noses*, somewhat hooked; *long curling hair*; little mouths; very fine, well-set, white teeth; full necks, and handsome breasts. They are very sharp and witty, and very talkative."

The Ashantí tribe of this race are among the most improved of the Pagan nations of Africa. Among them, as we are assured by Bowdich, are to be seen, especially "*among the higher orders*, not only the finest figures, but in many instances regular Grecian features, with brilliant eyes, set rather obliquely in the head. The women are rather like Indians than Africans."

[The Ashantí is the principal tribe of the Ochi race, which inhabits the woody and mountainous region westward of the Rio Volta, bounded by the sea on the south. According to the learned H. N. Riis, there are three tribes of this race, speaking mutually-intelligible dialects:—the Ashantís to the north, the Fantís to the south, and a tribe made up of three subdivisions, on the north-east, called Akim, Akwapim, and Akwam, here named in the order of succession from west to north-east. The writer's observations are chiefly drawn from experience among the Akwapims; but he intimates that they are applicable to the whole race.* He says the Akwapims have seventeen towns, each governed by its own chief, who has under him a council of elders, comprising the war-chief, the treasurer, the orator, the envoy, &c., who meet regularly in the town to administer the affairs of the state, to judge and decide the disputes of individuals, and to punish offenders. All the chiefs owe allegiance to the chief of Akropong, the head of the republic, whose relative position to his vassal chiefs is determined by the greater or less degree of energy which he displays in support of his prerogatives. All these offices are hereditary, but with the curious rule which subsists on the Coramandel coast also, that the son of a sister invariably succeeds to the dignity. The people are all cultivators, and they usually reside on their plantations, where they keep their numerous wives, children, and slaves. The whole population, according to a census taken by the English government, which succeeded that of Denmark, amounts to 11,442 souls, but M. Riis has doubts of its accuracy. The largest town, Abude, has 2,241; the chief place, Akropong, only 1,519.

The people believe in a God, who is the creator of the sun and of all things, and who made "the week of seven days,"† and to whom good men go after death. But this God leaves the government of the world to subordinate beings, who live in the mountains and valleys, the woods

* Elemente des Akwapim-Dialekts der Odschi-Sprache, &c. von. H. N. Riis, Basel, 1853. In Introduction, p. 4, the name of the race is said to be Otji or Tji (our Ochi or Chi), but the Author has been induced to prefer Odschi, because the sound of Ochi is "difficult to German ears and tongue."—ED.

† It would be interesting to know if this is an ancient tradition, or merely derived from the Mahometans, who live at no great distance. We are inclined to believe in its antiquity.—ED.

and waters, and who are invisible, though able to make themselves visible to the priests. Some of these beings are good, others malevolent, and the people lay in their way offerings of eggs and fruit, to show their gratitude to the former and to propitiate the latter. Fear is the grand motive power of the Negro-faith, and the multiplied superstitious practices, fetishes, charms, and horrible human sacrifices, so common in Africa, all are induced by this cause.

The Ochis declare they are immigrants in their present seats from a far northern land; two other languages are spoken by small tribes among them, perhaps the older population; but these tribes speak Ochi also.]*

The physical characters of this race will be illustrated by the outlines of skulls inserted in this and the next page. The first is the skull of a warrior of Ashantí; a cranium well formed, but somewhat shorter in the transverse diameter than the European. The arch of the forehead is somewhat low, and the ridge indicating the insertion of the temporal muscle strongly marked. The nasal bones are not so flat as in many African skulls. The zygoma is strong, and arched forward, not much outward, a characteristic of the prognathous skull, as distinguished from the pyramidal. One very remarkable character is that already alluded to in the general account of the peculiarities of the African cranium. The sphenoidal bone fails to reach the parietal bones, so that the coronal suture, instead of impinging upon the sphenoidal, as it does in most European heads, and in the human cranium in general, joins the margin of the temporal bone.

FIG. 73.

Skull of an Ashanti.

* Riis, Introduction, pp. v—viii.

FIG. 74.

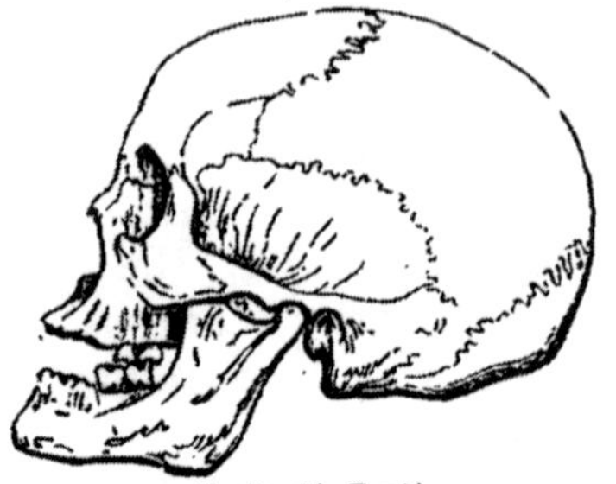

Skull of a Fanti.

The next is a figure of a skull of a native of the Gold Coast, probably of the Fantí race. It is described as generally of elongate form, with a slight longitudinal coronal eminence; the parietal bones bulge out, giving considerable breadth to the posterior part of the head. The forehead is narrow, but rises well; the nasal bones are broad and quite flat; the interorbital space considerable; the malar bones are large and prominent; the alveoli of the maxillary bone project obliquely forward, which is a strong characteristic of the prognathous skulls; the ascendant ramus of the lower jaw is broad, the posterior angle acute and prominent. The sphenoidal bone is in this skull united to the parietal, as usual.*

A district of the Gold Coast, of no great extent, around Acra, is inhabited by a particular race, speaking a language akin to that of the mountain Negroes of Adampi.† These people were described by the Danish missionary Isert. They have woolly hair, which is sometimes drawn out to the length of a half yard; it is generally black, but sometimes *red.*‡

---

* Martin's "Natural History of Mammiferous Animals," p. 297.

† The learned H. N. Riis, in p. v. of the Introduction to his Ochi Grammar, states that the Acra language is allied to the Ochi in the same degree as Swedish to German. The Editor must take exception to this remark; his objections are founded on some acquaintance with the Acra language, derived from Rask's Grammar, and Hanson's and Zimmermann's Versions of the Gospels, and from much intercourse with the Rev. A. W. Hanson, a native of Acra, an educated man and good philologer, who is also acquainted with Ashanti, which he speaks with fluency. The two languages have borrowed words from each other; but the grammatical structure differs considerably. The Acra people call themselves Gá in their own language.—ED.

‡ Dr. R. G. Latham has communicated a note to the Editor in reference

Farther eastward, the Slave Coast reaches the Rio Volta, and beyond that limit is the Bight of Benin, and still farther that of Biafra. The country behind the Slave Coast is inhabited by people of one race, to which belong the Negroes of Whidah, Ardrah, and Popo, as well as those of Dahomeh, a powerful and warlike tribe, in the interior. The natives of this country are tall and active men. In Benin, and to the south-eastward, are the races of Benin, Moko, and many others. Edwards says, that the Benins are of a yellowish-black colour, and that the shape of the face in most of them resembles that of the baboon, the lower jaw being more elongated than in the skull of any other Africans. I have examined the skull of a native of Benin, which had the Negro characters, but not in an extreme degree.

The region which I have last mentioned has been the great seat of the exportation of Negro slaves, and the tribes on the coast have been reduced to the lowest state of physical and moral degradation by the calamities and vices attendant on that traffic. Throughout Negroland, and especially this part of it, the inhabitants of one district in the interior, the dwellers on one mountain, are ever on the watch to seize the wives and children of the neighbouring clans, and to sell them to strangers: many sell their own. Every recess, and almost every retired corner of the land, has been the scene of hateful rapine and slaughter, not to be excused or palliated by the spirit of warfare, but

FIG. 75.

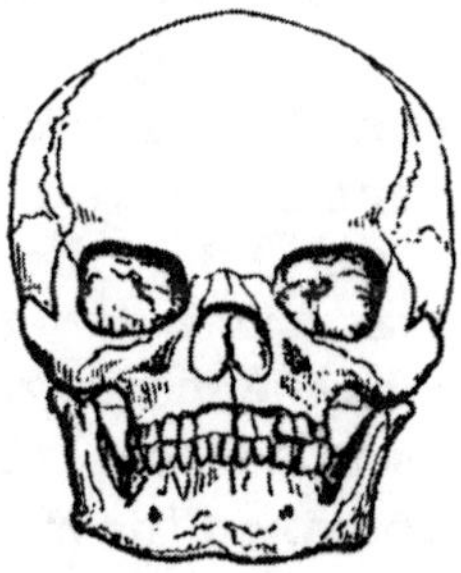

Skull of a Negro.

to the colour of the hair, observing that there is none so black as not to have a shade of red or auburn when seen in a cross light. He also notices the widely diffused habit of using alkaline washes—lime or soda—for the hair, which gives a red tinge, and has occasionally led to erroneous conclusions. This note is, perhaps, also applicable to some other statements in this work.—ED.

perpetrated in cold blood, and for the love of gain, and in accordance with laws deliberately enacted by parliaments for the professed purpose of enriching the merchants of enlightened and Christian Europe.

The preceding sketch gives a front view of a Negro skull, of similar character to those above mentioned.

## CHAPTER XVI.

### SOUTH AFRICAN NATIONS.

ALTHOUGH that part of Africa which is to the southward of the equator, as well as the central region of the continent to the northward of the same line, contains vast spaces which have not yet been penetrated by European travellers, our knowledge respecting the country and its inhabitants has been considerably increased during the last few years. This is partly the result of expeditions of discovery undertaken by order, or with the sanction, of the Colonial Government at the Cape of Good Hope, by persons better qualified for the purpose by previous acquirements than most former travellers had been, and more amply provided with the means requisite to ensure success.* A new light, also, has been thrown over a wider extent of Southern Africa, on the mutual relation of the native tribes, by missionaries, who, in the hope of preparing a way for the propagation of Christianity among the aboriginal inhabitants, have laboured successfully to make themselves fami-

* See a "Report of the Expedition for Exploring Central Africa from the Cape of Good Hope, June 23, 1834, under the superintendence of Dr. Andrew Smith," Cape Town, 1836. Also, "Report of an Expedition of Discovery through the Countries of the Great Namáquas, Boschmans, and the Hill Dámaras, in South Africa," by Captain J. E. Alexander, R.L.S. &c. (Journal of the Royal Geographical Society, vol. viii.)

Many interesting publications may be added to the list since the above note was written by the Author.—ED.

liar with several of their languages. These meritorious persons have also achieved what is even more important with a view to the extension of knowledge, and which may hereafter prove to be of still greater and more permanent value with relation to objects of the highest interest:—they have accomplished the arduous task of reducing to writing, and subjecting to grammatical analysis, the intricate and difficult combinations of these languages, which, in the complexity and almost infinite variety of their inflexions, may be compared to the laboriously-constructed idioms of the North American nations. The result of these researches into the glossology of Southern Africa is similar to that which has been attained from inquiries of the same description in regard to the languages of other barbarous countries. In all such regions—as, for example, in America and in Australia—it was long believed that the native population consisted of an immense number of distinct aboriginal tribes. The inhabitants of almost every valley, or separate plain, or mountainous tract, were supposed to have a language of their own, unconnected with the idioms of their nearest neighbours. Wherever sufficient inquiry has been made, a more accurate acquaintance with facts has proved the fallacy of this opinion, and has shown that a few mother tongues, mostly divided into a variety of dialects, are spread over vast spaces. In proportion as the inquiry has been more accurately pursued, and a scientific examination of languages has advanced, in the same degree the number of languages supposed to be distinct has been from time to time diminished; with the number of separate languages, that of distinct races, or families of nations, has been in proportion reduced. These observations are, perhaps, in no instance more applicable than they are to the languages and nations of Africa. If we survey the languages in reference to the present state of our knowledge, we may perhaps venture to say that three-fourths of the whole extent of this Continent are occupied by three great families of nations. From these regions we exclude the whole of Central Negroland or Soudan, meaning the countries within the tropics and to the northward of the equator, or immediately under that line, where Africa has its greatest breadth, and throws out towards the Atlantic the great projecting land of Senegambia and Guinea. The woolly-haired tribes who inhabit Central Negroland, and Senegambia, and Guinea, cannot properly be termed a family of nations; because, although they bear to each other a striking physical resemblance, we have as yet no proof, founded on an analysis of their languages, of a common link between them, or of their descent from a common stock.* If such a connexion should hereafter be found to exist, we shall then be enabled to refer nearly all the African nations to four great families. The three principal groups which, to the exclusion of the last-mentioned Negro nations, occupy nearly three-fourths of Africa, are the following:—The first are races more or less nearly allied to the Syro-Arabian family: this comprehends the Lybian or Atlantic nations in the north, and the Abyssinian races to the east,—people who are spread over nearly all the north of Africa, from Mount Atlas to the Arabian Gulf. The second family are the Galla and the tribes related to them in the east and to the southward of Abyssinia. It seems not improbable that these last nations, viz. the Galla, may ultimately be found referable to the former group. The third is, perhaps, a still more widely dispersed family, occupying, we have reason to believe, nearly all that vast region of Africa which reaches from the equator to the southern tropic, as well as a considerable part of the territory that intervenes between the same line and the southern extremity of the continent. To this family of nations I shall apply the distinctive epithet of the Great South African Race. They do not, however, occupy the whole of Africa beyond the southern tropic. A great tract of country between the Orange River and the Cape of Good Hope has been the abode from immemorial time of a race of very different physical

* This remark may now be modified. See notes in pp. 323 and 334.—ED.

and moral character. They are the Lesser Race of Southern Africans. Their character and history have been elucidated by late researches; and the investigation will lead us to some facts which have a bearing on the origin and connexions of various nations, not only in Africa, but likewise in several other parts of the world.

The following pages will be devoted to an attempt to throw some light on the ethnography and history of these two southern groups of nations, or to collect and systematise the scattered notices that are to be found in various quarters respecting them, some of which have lately come to my knowledge. I shall begin with the Lesser South African Race.

END OF VOL. I.